Report no. VNTSC-TFM-08-09

Characterization of Uncertainty in ETMS Flight Events Predictions and its Effect on Traffic Demand Predictions

Eugene P. Gilbo
Scott B. Smith

July 11, 2008

Prepared for
Federal Aviation Administration
Office of System Operations Programs
600 Independence Avenue, SW
Washington, DC 20591

Prepared by:
Volpe National Transportation Systems Center
Traffic Flow Management Division
55 Broadway
Cambridge, MA 02142

Acknowledgments

The authors would like to thank Midori Tanino and Ved Sud of the FAA for their support and encouragement. Within the Volpe Center, we would like to thank Dick Bair, Rick Oiesen and Ken Howard for reviewing the report and their valuable comments and suggestions that improved its content. We also thank Mary Costello for editing this report.

Executive Summary

The Enhanced Traffic Management System (ETMS) predicts traffic demand in the National Airspace System (NAS) up to 24 hours in the future to determine potential congestion in airspace or airports. To identify congestion, it finds time intervals and NAS elements (i.e., sectors, airports, and fixes) where predicted demand exceeds the capacity that has been input into ETMS. Based on the duration and magnitude of congestion, traffic flow management (TFM) specialists decide whether to take action to bring traffic demand down to capacity through various traffic management initiatives (*TMI*s), such as Ground Delay Programs (GDPs), Airspace Flow Programs (AFPs), or Miles-in-Trail (MIT). ETMS produces deterministic predictions of traffic demand and does not take into account the random errors in these predictions. This uncertainty in predictions creates uncertainty in the information that TFM specialists use in their decision-making process.

A recent direction in TFM research is concerned with acknowledging the uncertainty in predictions and creating probabilistic TFM that considers the uncertainty in the decision-making process. The premise behind probabilistic TFM is that traffic managers will make better decisions if they use data and tools that reflect the uncertainty in the system. Probabilistic TFM is based on probabilistic representation of traffic demand and capacity of NAS elements through the respective probability distributions that allow for determining the probabilities of congestion. In order to obtain the probability distributions for probabilistic TFM, a thorough statistical analysis of prediction errors is needed to characterize the prediction uncertainties.

This report presents the results of analysis and characterization of uncertainty in traffic demand predictions using ETMS data and probabilistic representation of the predictions. Our previous research, described in two prior reports, was focused on analysis of aggregate 15-minute traffic demand predictions in ETMS, on improving the accuracy of these predictions and increasing the stability of the ETMS monitor/alert function, while not explicitly considering the uncertainty in predictions of flight events for individual flights. This study continues the previous one. It also focuses on uncertainty in traffic demand predictions, but, unlike the previous one, it explicitly considers uncertainty in individual flights' predictions for estimation of uncertainty in aggregate demand count predictions at NAS elements and for probabilistic representation of those predictions.

The major steps in this study include

- Characterization of uncertainty in predicting times for individual flights via statistical analysis of prediction errors
- Estimation of predicted probabilities for individual flights to be in a NAS element at any time by using the probability distribution functions of prediction errors
- Estimation of probability distributions of predicted 15-minute demand counts for NAS elements by using the predicted probabilities for individual flights to be in the NAS element (in this report, for airport arrivals)
- Probabilistic prediction of 15-minute traffic demand counts that includes expected value of traffic demand and a range of uncertainty around the expected demand with a certain probability for the predicted demand to be within this range
- Comparison of probabilistic and deterministic demand count predictions.

The study is based on statistical analysis of historical ETMS data on approximately 30,000 flights for nine arrival airports and 13 en route sectors during four days of 2007: two days in April (April 25 and 26) and two days in June (June 7 and 8).

Statistical analyses and characterization of uncertainty in predicting times for individual flights were focused on prediction errors in flights' arrival time at airports and time of flights' crossing sector boundaries. The analysis was conducted for various look-ahead times (LATs) up to three hours in the future, and the characteristics of prediction errors were separately estimated for active and for proposed flights that have not departed from origin airports.

Major findings from this analysis include the following:

- Sector entry time and airport arrival time predictions for active flights are much more accurate than for proposed flights. For active flights, the standard deviation of arrival time error was in the 10 minute range, while for proposed flights this standard deviation would typically be in the 25-minute range.

- For active flights, the distributions for prediction errors of airport arrival or sector entry times were either symmetric or were slightly asymmetric with a higher probability for active flights to arrive at an airport or to enter a sector earlier than predicted

- For proposed flights, the distributions were asymmetric the other way, with a higher probability for a flight to arrive later than predicted.

- The prediction error for time-in-sector was noticeably smaller than for the sector entry time. The standard deviation of time-in-sector error is typically between 3 and 6 minutes (depending on the sector's size) whereas the standard deviation of the error of sector entry time is between 5 and 32 minutes (depending on flight status).

- The probability that a given flight will actually be in the sector during the time that it was predicted to be in the sector depends heavily on both the in-sector time and the flight status (active or proposed). For example, when the predicted in-sector time was less than 20 minutes, fewer than 50% of proposed flights predicted to be in a sector were actually in the sector at some point during the predicted interval.

The results of the analysis of uncertainty in individual flights predictions were used to develop probabilistic predictions of aggregate traffic demand counts. A methodology was developed to determine the probability distribution of predicted aggregate 15-minute demand counts at arrival airports by using the probability distributions of errors in arrival time predictions for individual flights. A key element of the methodology is that the probability distributions of traffic demand counts for a 15-minute interval were estimated from an extended set of candidate flights that included, along the flights predicted to arrive at the interval of interest, also flights with estimated times of arrival in adjacent 15-minute intervals. The probability distributions of traffic demand counts are used for probabilistic predictions of traffic demand that include the expected values of traffic demand and the range of uncertainty around those expected values with a certain probability for the predicted demand to be within the area.

A simplified analytical method is proposed to estimate expected values of traffic demand counts. As a result, the expected value of the 15-minute count for an interval was found to depend on forecasts of demand counts for adjacent intervals, as well as the interval of interest, and is equal to a weighted sum of those predicted demand counts. The coefficients appeared remarkably similar to those derived from regression analysis of 15-minute count data in the previous study.

Numerical examples illustrated the difference between deterministic and probabilistic traffic demand predictions.

Table of Contents

1. Introduction 1
2. Examined Flight-by-Flight Data 3
3. **Analysis of ETMS Predictions of Individual Flight Events** 5
    3.1 Sectors 5
        3.1.1 Initial Analysis 5
        3.1.2 Errors in Predicted Sector Entry Times 10
    3.2 Airports 13
4. **Probability that a Flight will be in a Sector** 21
    4.1 Time in Sector 21
    4.2 Probability that a Flight will be in a Sector 23
5. **Probabilistic Count Predictions** 27
    5.1 Introduction 27
    5.2 Probability that a Flight will Arrive During a Specified Interval 27
    5.3 Combining Probabilistic Count Predictions 31
    5.4 A Tractable Probabilistic Count Prediction 32
    5.5 Comparison with Previous Work 42
6. **Conclusion** 44
Appendix A   Flight Event Predictions and 15-minute Counts: Simulation Experiments . 46
Appendix B   References 50

1. Introduction

This report is a continuation of the research on improving accuracy and reducing uncertainty in Enhanced Traffic Management System (ETMS) traffic demand predictions. Previous research (see [2] and [4]) was focused on analysis of accuracy of 15-minute aggregate traffic demand predictions in ETMS and on the development of a regression model aimed at improving the accuracy and stability of those aggregate predictions. The regression model improved both the accuracy of demand predictions and the stability and accuracy of ETMS Monitor/Alert.

Many factors contribute to uncertainty in predictions of aggregate traffic demand. Although the major contributor into uncertainty in aggregate demand predictions is uncertainty in predicting flight events such as sector boundary crossings or arrival times at airports for individual flights, the characteristics of this kind of uncertainty were not explicitly taken into account in our previous research on aggregate demand predictions. Taking into account uncertainty in predictions of the times for individual flights to be in National Airspace System (NAS) elements along the flight's origin-destination routes would permit probabilistic predictions of aggregate traffic demand, which, in turn, would improve probabilistic Traffic Flow Management (TFM) decision-making [5, 6].

The purpose of this report is to

- Analyze the errors in time predictions for individual flights and characterize the accuracy of ETMS flight-by-flight prediction data
- Use these characteristics of uncertainty to determine the probability for an aircraft to be in a NAS element at any given time (e.g., to be in sector or to arrive at an airport)
- Use these probabilities of individual flights' events for probabilistic predictions of aggregate traffic demand counts at NAS elements (e.g., 15-minute demand counts at airports).

Larry Meyn, in his pioneering publication [3], proposed a methodology for obtaining probabilistic aggregate traffic demand prediction through characteristics of uncertainty in individual flight predictions. In [3], a recurrent algorithm was presented for calculating the probability distribution of predicted aggregate traffic demand based on the probabilities of individual flights to be in the time interval of interest (e.g., 15-minute interval). It is worth mentioning that in [3], the set of candidate flights for probabilistic distribution of predicted aggregate demand included only the flights with the ETAs within the time interval of interest and did not include the flights with ETAs outside this interval. To estimate the benefits of probabilistic demand predictions over deterministic ones, a Monte Carlo simulation was used in [3]. The simulation was conducted on artificial data (not actual air traffic data) and showed that the probabilistic approach provided more accurate predictions than the deterministic approach. Some elements of the methodology from [3] are used in this study. Unlike [3], this study is based on ETMS historical data. Extensive statistical analysis of ETMS data was performed to characterize the accuracy of predictions of individual flight events. Those characteristics were then used for probabilistic predictions of aggregate traffic demand for specific time intervals. The extended set of candidate flights for those predictions included the flights with ETAs both within the time interval of interest and outside the interval, namely, in several adjacent earlier and later intervals. The report presents a methodology and the corresponding techniques of how to use the data for probabilistic prediction of aggregate traffic demand at airports and at sector boundary crossing.

The report has been organized as follows.

- Section 2 describes the data that was used in this study.
- Section 3 examines the errors that ETMS makes when predicting sector entry or airport landing time. It summarizes these errors with empirical distributions.

- Section 4 presents an empirical analysis of the probability that a flight will be in a sector at a particular moment, given predicted entry time and predicted time in sector.
- Section 5 uses the probability distributions of errors in predictions of individual flight events (from Section 3) to derive probability distributions of the numbers of arrivals in a NAS element (airport or sector boundary) in a 15-minute interval. From this, probabilistic demand predictions for airport arrivals that comprise the mean demand and the area of uncertainty around the mean demand are obtained.
- Section 6 concludes the report by summarizing the findings.
- Appendix A presents a simulation model that was developed to examine the relationship between the accuracy of flight event predictions and count predictions.

2. Examined Flight-by-Flight Data

ETMS continuously updates information on the status of each flight in the system and predicts each flight's time and location at various points along its origin-destination route. The ETMS flight list for a NAS element can be requested at any time. The flight list shows the flight's status (airborne or still on the ground), estimated time of departure (or actual departure time for an active flight), and estimated time of arrival at a NAS element, including destination airport. ETMS also collects flight-by-flight historical data that include both predictions and what actually happened.

On April 25 and 26, 2007, and again on June 7 and 8, 2007, ETMS list requests were repeatedly run for the following nine airports and thirteen en-route sectors (Figure 2-1):

- ORD, ATL, DFW, LAX, MIA, BOS, SFO, STL, MCI
- ZBW02, ZBW17, ZID82, ZID83, ZID86, ZLC06, ZLC16, ZMP20, ZOB57, ZOB67, ZOB77, ZSE14, ZTL43

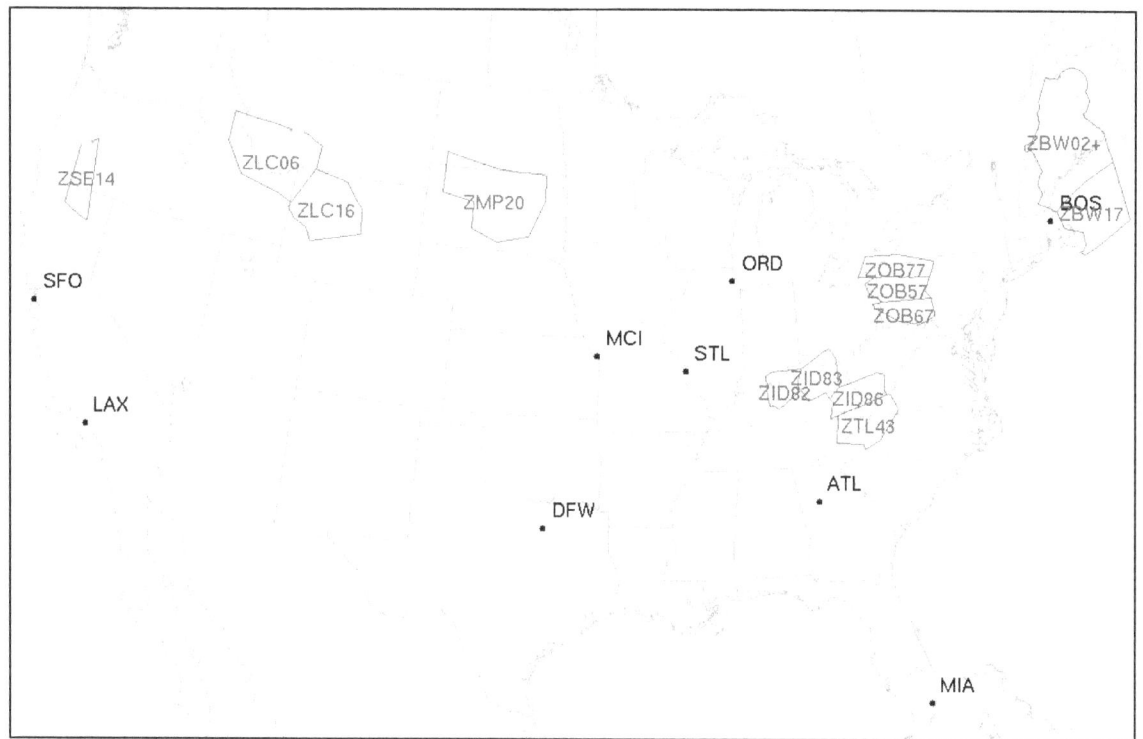

Figure 2-1 Examined Airports and Sectors

These repeated list requests, which were run once every 15 minutes, generated flight-by-flight predictions for airport arrival, sector entry and sector exit. Data was gathered for approximately 7 hours on the afternoons of April 25 and 26, 2007 and for approximately 12 hours on the afternoons of June 7 and 8. April 25th was a Wednesday and June 7th was a Thursday.

For the airports on April 25-26, there were some 6,700 flights, with a total of 78,000 observations. (Each flight has multiple observations because predictions are made at various look-ahead times.) On June 7-8, there were approximately 6,000 flights, with a total of 93,000 observations. For the sectors on April 25-26, there were approximately 12,000 flights, with 125,000 observations. On June 7-8, there were approximately 13,000 flights, with 150,000 observations.

Statistical analysis of flight data was performed to characterize uncertainty (errors) in flight event predictions. Separate analysis was conducted for active (airborne) flights and for proposed flights (that are still on the ground) to characterize the difference in the accuracy of predictions depending on the flight's status.

3. Analysis of ETMS Predictions of Individual Flight Events

This section provides a descriptive statistical analysis of the accuracy of ETMS predictions that will be used in later sections. The primary figure of merit used is the accuracy of predicted flight arrival times at a NAS element, which is measured as an error (predicted – actual) in the arrival time. For the sector data, this is the (predicted – actual) sector entry time, although prediction accuracy of time-in-sector is also considered.

Separate statistical data analyses were conducted for airports and sectors, although the steps taken were similar:

- Initial analysis (exploratory data analysis) and removal of outliers
- Statistical estimation of parameters that characterize distributions of prediction errors (mean, standard deviation, skewness)

3.1 Sectors

3.1.1 Initial Analysis

Table 3-1 shows the distributions of (predicted – actual) sector entry time, divided into the following time buckets.

- Less than -180 minutes (flights more than 3 hours late)
- -180 to -61 minutes (flights 1 – 3 hours late)
- -60 to -15 minutes (flights less than 1 hour late)
- On time (flights that are between 15 minutes late and 15 minutes early)
- 15 – 60 minutes (flights less than 1 hour early)
- 61 – 180 minutes (flights 1 – 3 hours early).

No flights were more than 3 hours early.

The LAT column shows Look-Ahead Time (LAT) range for predictions. In the Flight Status column, letter A denotes active, i.e., airborne flights, and letter P denotes the proposed flights that are still on the ground at origin airports. The results in this table are not surprising. Flights in the air tended to be early, while flights on the ground tended to be late due to departure delays. For example, Table 3-1 shows that approximately 2 to 24 percent of active flights enter sectors minutes earlier than predicted and only 1 to 3 percent of active flights enter sectors later than predicted. However, more than 30% of the flights still on the ground cross sector entry boundaries later than predicted, and roughly 6 to 10% enter sectors earlier than predicted. Note that "early" and "late" here refers to the sign (positive or negative) of the (predicted – actual) time, and is related to the latest known flight plan. For example, if the original (schedule) flight plan resulted in estimated sector entry time of 1300Z, but its origin airport departure was delayed by two hours so that its updated flight plan results in a new estimated sector entry time of 1500Z and the flight actually enters the sector at 1435Z, it would show up in this table as "early," even though according to the original schedule, the flight is nearly two hours late.

Table 3-1 Sectors: Distribution of Entry Time Errors

Month	LAT	Flight Status	Number of Observations	Late (minutes)			On Time[1]	Early (minutes)	
				Over 180	61 – 180	15 - 60		15 - 60	61-180
April	1-2 hr	A	3194	0.1%	0.2%	1.0%	94.7%	3.7%	0.2%
June	1-2 hr	A	5690	0.0%	0.2%	1.6%	96.5%	1.6%	0.1%
April	2-3 hr	A	1853	0.0%	0.0%	2.9%	80.2%	16.2%	0.6%
June	2-3 hr	A	4204	0.0%	0.0%	1.8%	74.6%	23.2%	0.3%
April	1-2 hr	P	10474	0.6%	3.9%	26.0%	63.1%	6.2%	0.2%
June	1-2 hr	P	10691	0.9%	5.8%	26.0%	60.9%	6.4%	0.1%
April	2-3 hr	P	4962	0.5%	5.3%	28.2%	56.5%	8.5%	0.9%
June	2-3 hr	P	5443	1.2%	7.3%	28.6%	52.5%	9.4%	1.0%

Figures 3-1 through 3-4 show the distributions of sector entry time predictions, with separate curves for April and June data, short (1-2 hr) and longer (2-3 hr) look-ahead times, and active (airborne) and proposed (still on the ground) flights. Both the probability densities (Figures 3-1 and 3-3) and cumulative distributions (Figures 3-2 and 3-4) are shown.

Figures 3-1 through 3-4 show that

- The range of prediction errors in sector entry times is significantly narrower for active flights than for proposed flights, confirming that predictions for airborne flights are more accurate than those for flights still on the ground.

- For active flights, the predictions made for 1 to 2 hours in the future are noticeably more accurate than the ones made for 2 to 3 hours in the future.

- For active flights, the distributions of prediction errors for a shorter LAT (from 1 to 2 hours) appear symmetrical with the average and median error close to zero, whereas for a longer LAT (from 2 to 3 hours) the distribution is asymmetric with a heavier right-hand tail and median error within 5 – 7 minutes, i.e., the distribution is biased toward earlier arrivals than predicted. This can be seen more clearly from Figure 3-2 and Figure 3-4 by comparing the values of errors that correspond to the probability 0.5 on the cumulative distribution curves and the behavior of the curves on the right and on the left from 0.5 probability.

- For the proposed flights, there is no significant difference in prediction accuracy between predictions for a shorter and a longer LAT. The distributions of prediction errors are asymmetric with heavy left-hand tails, which reflect the tendency for flights on the ground to, on average, enter sectors later than predicted.

- The probability density plots (Figures 3-1 and 3-3) show a number of outliers (the "bumps" in the plots at values far from zero).

---

[1] "On-time" is considered to be within 15 minutes (early or late) of the predicted time.

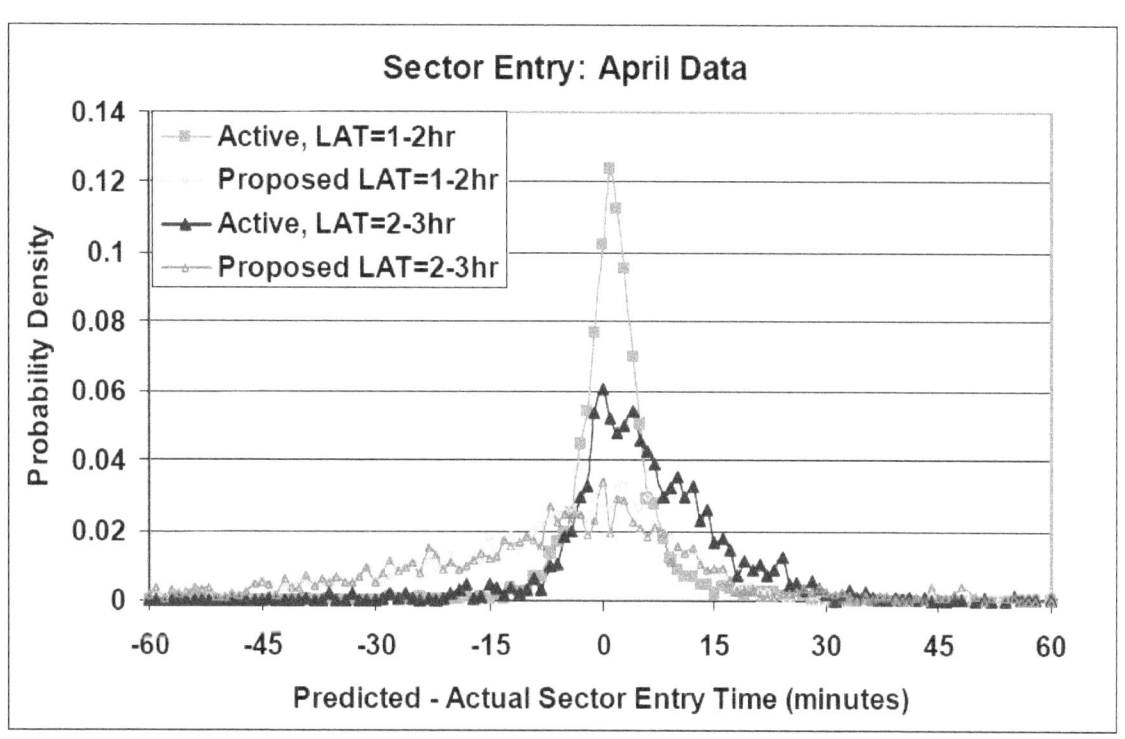

Figure 3-1 Distribution of Errors in Sector Entry Time Predictions: April Data

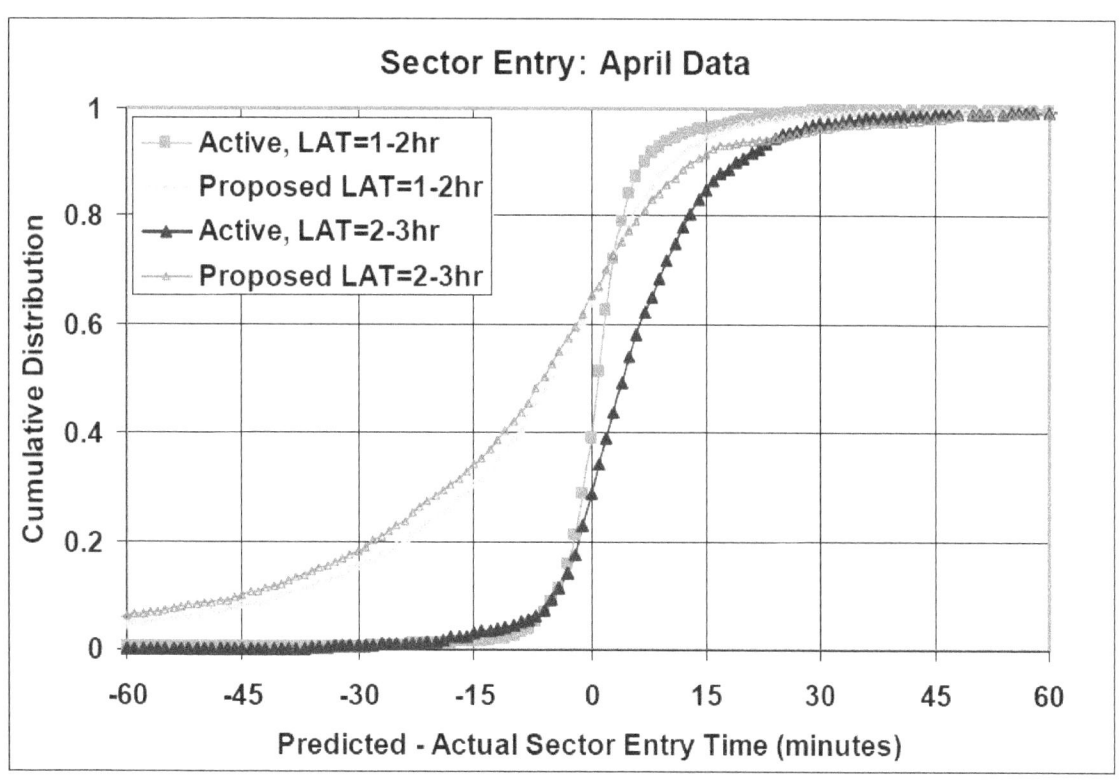

Figure 3-2 Cumulative Distribution of Errors in Sector Entry Time Predictions: April Data

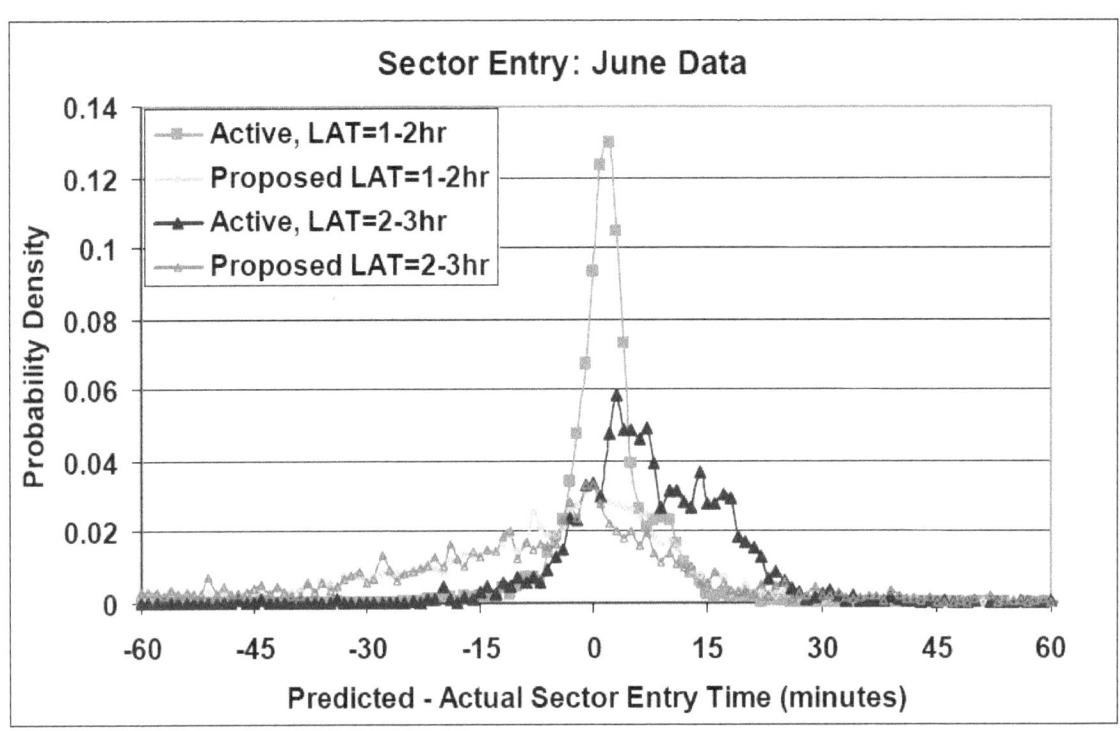

Figure 3-3 Distribution of Errors in Sector Entry Time Predictions: June Data

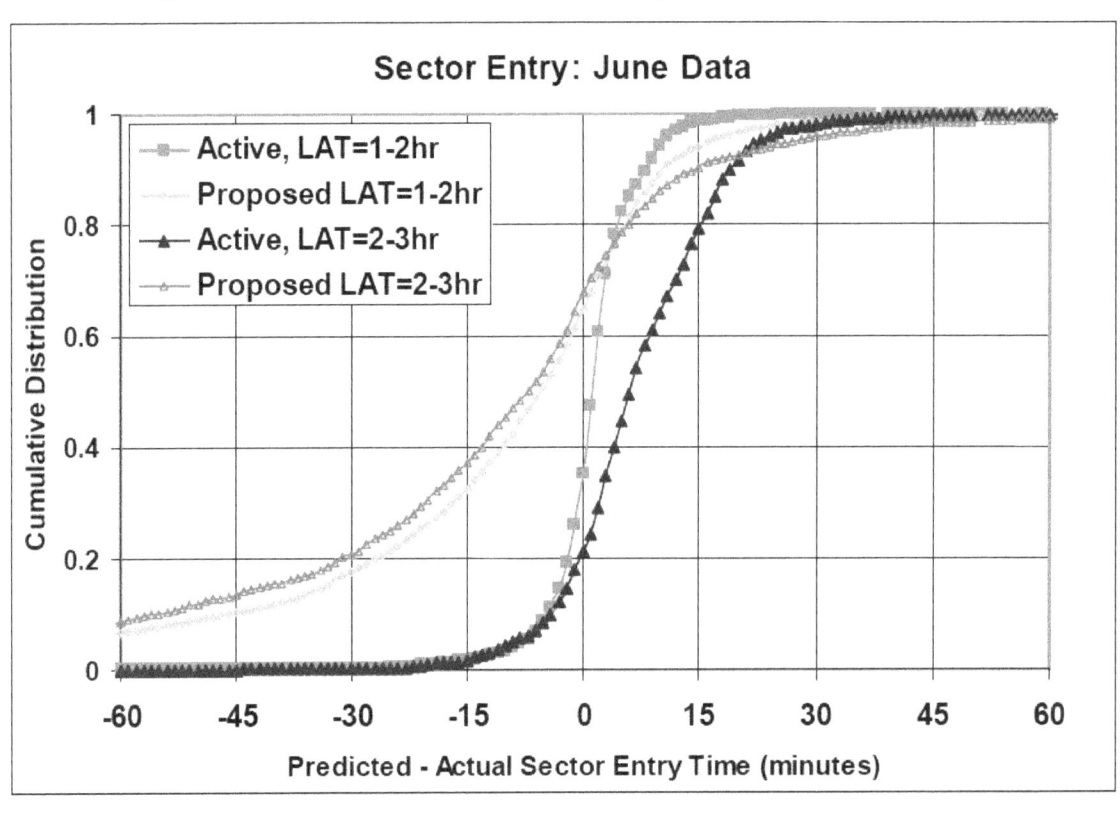

Figure 3-4 Cumulative Distribution of Errors in Sector Entry Time Predictions: June Data

Figures 3-5 and 3-6 show both the probability density and cumulative distribution for the combined data with a look-ahead time of 1 to 2 hours. The distribution is asymmetric with a heavy left-hand tail with the shape similar to the distribution of errors for proposed flights. (Flights tend to enter sectors later than predicted). They show that when the data for active and proposed flights are combined in a single set, the inaccuracy in predictions for proposed flights prevailed and determined the shape of the probability distribution.

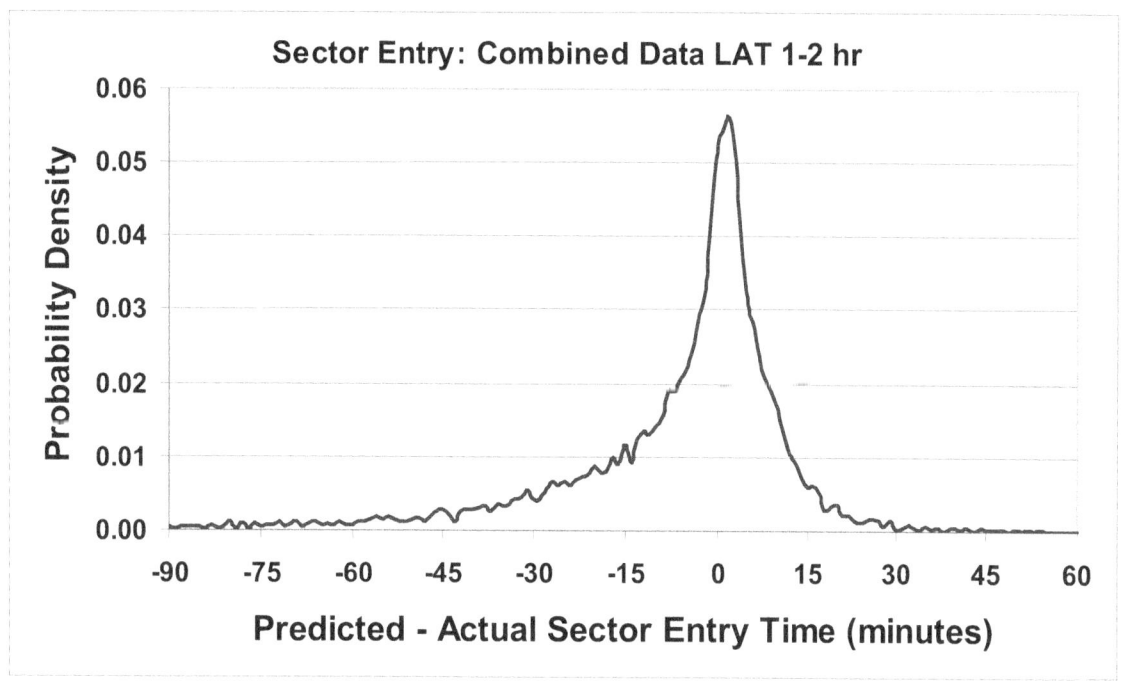

Figure 3-5 Distribution of Errors in Sector Entry Time Predictions: Combined Data

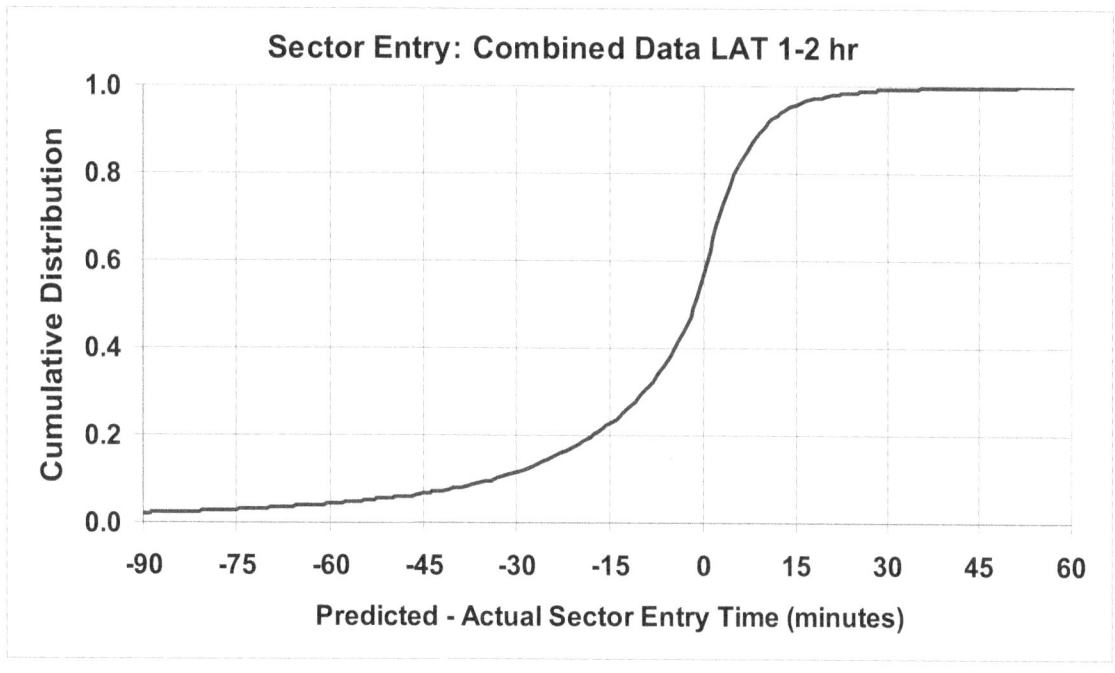

Figure 3-6 Cumulative Distribution of Errors in Sector Entry Time Predictions: Combined Data

A number of the extremely late flights were individually examined. Most of the late flights were general aviation flights. Several were commercial flights. Three of the commercial flights were checked in the Bureau of Transportation Statistics flight delay database. All three flights also showed delays of at least several hours in that database.

3.1.2 Errors in Predicted Sector Entry Times

The results of statistical data analysis are presented in Tables 3-2 through 3-5. For the analysis, the data was used only for the flights with no more than 3-hour late entering sectors. The data for the flights more than 3 hours later or early were considered to be outliers and were removed from the collected historical data sets. Recall from Table 3-1 that such flights represent only a small portion (less than 1% overall) of the total. Once the outliers were removed, the average, standard deviation and skewness of prediction errors were estimated.

The standard deviation characterizes the spread of data about the mean. In normally distributed data, some 68% of the observations will fall within one standard deviation of the mean, and approximately 95% of the observations will fall within two standard deviations of the mean.

The skewness characterizes an asymmetry of the distribution relative to its mean. Negative skewness indicates that the left tail of the distribution is longer and heavier than the right tail. In our case, it means that the probability for a flight to arrive later than predicted is higher than the probability to arrive earlier. (For large n, skewness is approximated as $(1/n) \sum ((x - Avg(x))/s)^3$ where n is the number of observations and s is the standard deviation. For a symmetric distribution such as the normal distribution, the skewness is 0.)

Tables 3-2 and 3-4 consistently show negative average errors in sector entry time predictions and negative estimates of skewness in the error distributions for the flights still on the ground (proposed flights) at all sectors analyzed. Negative average errors indicate that the proposed flights enter sectors, on average, later than predicted. Negative skewness indicates that the probability distributions of prediction errors for those flights are asymmetric with heavier left-hand tail. The average errors, standard deviations of the errors and skewness vary for various sectors, but the values of the estimates vary within relatively narrow ranges (with the exception of one or two sectors).

For active flights, Tables 3-3 and 3-5 consistently show positive average errors in sector entry time predictions (airborne flights enter sectors, on average, earlier than predicted). The average prediction errors vary from one sector to another but the errors are significantly smaller than absolute prediction errors for the non-active flights presented in Tables 3-2 and 3-4. Standard deviations of prediction errors for the active flights are significantly (and consistently) smaller than for the inactive flights (predictions for active flights are more accurate than for proposed flights). Tables 3-2 through 3-5 illustrate the magnitude of the differences in prediction accuracy for those two types of flights.

Table 3-2 April Data, Flights on the Ground, (Predicted-Actual) Sector Entry Time (minutes)

LAT	Sector	Number of Observations	Average Error	Standard Deviation	Skewness
1 – 2 hr	ZBW02	774	-10.16	24.4	-1.8
	ZBW09	1282	-4.94	21.4	-1.7
	ZBW17	773	-14.74	22.8	-1.1
	ZBW20	2827	-11.79	24.8	-2.2
	ZBW46	1562	-7.44	24.9	-2.3
	ZMP20	854	-17.81	28.6	-2.3
	ZOB57	1516	-9.92	24.2	-2.1
	ZOB67	2135	-12.79	24.4	-1.7
	ZOB77	1390	-10.68	22.3	-2.3
2 - 3 hr	ZBW02	676	-12.66	22.6	-0.5
	ZBW09	1054	-5.53	24.7	-2.4
	ZBW17	785	-11.35	22.5	-1.3
	ZBW20	2441	-12.10	25.5	-2.1
	ZBW46	1292	-8.24	26.2	-2.0
	ZMP20	1127	-12.61	31.2	-2.3
	ZOB57	1169	-11.32	25.3	-1.7
	ZOB67	1799	-12.68	26.4	-1.7
	ZOB77	1230	-9.32	20.8	-1.1

Table 3-3 April Data, Flights in the Air, (Predicted-Actual) Sector Entry Time (minutes)

LAT	Sector	Number of Observations	Average Error (min)	Standard Deviation	Skewness
1 – 2 hr	ZBW02	851	4.22	6.4	1.3
	ZBW09	192	1.32	7.0	0.4
	ZBW17	226	2.28	8.5	1.4
	ZBW20	641	0.05	4.5	-3.3
	ZBW46	205	5.19	8.8	2.3
	ZMP20	803	-1.69	12.6	-7.8
	ZOB57	76	3.26	7.0	-1.0
	ZOB67	236	1.50	8.6	-7.2
	ZOB77	291	1.16	7.6	-8.1
2 – 3 hr	ZBW02	819	9.99	10.9	0.6
	ZBW09	87	1.10	12.1	0.7

LAT	Sector	Number of Observations	Average Error (min)	Standard Deviation	Skewness
	ZBW17	103	8.95	15.1	0.3
	ZBW20	489	2.07	8.4	-1.1
	ZBW46	141	6.77	9.2	0.1
	ZMP20	234	3.45	10.0	0.8
	ZOB57	12	12.21	21.2	-1.5
	ZOB67	135	2.15	7.5	0.0
	ZOB77	66	2.76	6.5	-0.4

Table 3-4 June Data, Flights on the Ground, (Predicted – Actual) Sector Entry Times

LAT	Sector	Number of Observations	Average Error (min)	Standard Deviation	Skewness
1 – 2 hr	ZBW02	1309	-14.55	27.5	-1.8
	ZBW09	1163	-12.09	32.2	-2.1
	ZBW17	530	-13.36	26.0	-2.3
	ZBW20	2962	-16.14	29.6	-2.2
	ZBW46	1688	-8.53	22.4	-1.8
	ZMP20	1318	-16.96	25.7	-1.7
	ZOB57	1449	-14.25	26.6	-1.2
	ZOB67	1856	-11.74	25.8	-1.9
	ZOB77	1400	-13.76	30.4	-2.3
2 – 3 hr	ZBW02	1216	-14.53	26.3	-1.3
	ZBW09	968	-10.19	30.9	-2.0
	ZBW17	549	-13.14	28.3	-1.9
	ZBW20	2506	-18.34	31.2	-2.1
	ZBW46	1436	-7.90	24.0	-1.9
	ZMP20	1676	-14.73	27.6	-2.0
	ZOB57	1147	-13.03	27.6	-1.1
	ZOB67	1517	-11.90	28.2	-1.6
	ZOB77	1124	-13.35	30.9	-2.2

Table 3-5 June Data, Flights in the Air, (Predicted-Actual) Sector Entry Times

LAT	Sector	Number of Observations	Average Error (min)	Standard Deviation	Skewness
1 – 2 hr	ZBW02	1389	2.70	4.7	0.6
	ZBW09	172	-0.65	5.5	-0.4
	ZBW17	1196	0.36	7.9	-2.5
	ZBW20	863	0.58	3.9	-1.6
	ZBW46	861	6.27	5.1	-0.6
	ZMP20	1082	-2.30	14.3	-8.4
	ZOB57	114	2.46	6.6	-0.5
	ZOB67	258	1.02	6.7	-2.7
	ZOB77	218	-0.59	12.2	-8.2
2 – 3 hr	ZBW02	1370	11.27	9.0	0.4
	ZBW09	96	1.77	6.8	1.6
	ZBW17	1145	3.90	11.5	-0.3
	ZBW20	654	4.71	8.9	0.1
	ZBW46	739	11.30	8.9	0.0
	ZMP20	296	2.74	13.9	-2.5
	ZOB57	19	4.37	5.2	1.3
	ZOB67	128	0.29	6.7	-1.1
	ZOB77	66	1.00	7.1	-2.0

3.2 Airports

This section presents the results of analysis of accuracy of flight arrival time predictions. The analyses were conducted on historical data for individual flights collected at nine airports during two days in April and two days in June, 2007. This section is organized in the following manner.

First, we present the results of statistical analysis that were derived from processing of the extended set of historical data, which consolidated the data from all nine airports considered in the study. The corresponding results are presented in Table 3-6 and Figure 3-8 through Figure 3-11. The results of statistical analysis of flight arrival data for each airport separately are then shown in Table 3-7 through Table 3-10.

Table 3-6 shows the distributions of (predicted – actual) airport arrival time, divided into the following time buckets:

- Less than -180 minutes (flights more than 3 hours late)
- -180 to -61 minutes (flights 1 – 3 hours late)
- -60 to -15 minutes (flights less than 1 hour late)
- On time (flights that are between 15 minutes late and 15 minutes early)

- 15 – 60 minutes (flights less than 1 hour early)
- 61 – 180 minutes (flights 1 – 3 hours early).

Table 3-6 Airports: Distribution of Errors in Flight Arrival Time Predictions

Month	LAT	Flight Status	Number of Observations	Time Bucket					
				Late (minutes)			On Time[2]	Early (minutes)		
				Over 180	61 – 180	15 - 60		15 - 60	61- 180	Over 180
April	1-2 hr	A	7476	0.0%	0.1%	4.4%	92.6%	2.0%	0.0%	0.9%
June	1-2 hr	A	8123	0.1%	0.2%	4.5%	92.7%	2.5%	0.0%	0.1%
April	2-3 hr	A	5699	0.0%	0.0%	4.5%	91.0%	3.9%	0.1%	0.4%
June	2-3 hr	A	4605	0.1%	0.2%	4.8%	89.7%	4.9%	0.3%	0.1%
April	1-2 hr	P	12605	0.2%	5.7%	25.6%	59.9%	6.9%	1.3%	0.4%
June	1-2 hr	P	7159	0.9%	10.6%	30.2%	53.4%	4.1%	0.2%	0.4%
April	2-3 hr	P	10490	0.3%	4.8%	25.3%	59.0%	9.0%	0.6%	1.0%
June	2-3 hr	P	12382	1.1%	10.4%	27.7%	52.8%	7.0%	0.6%	0.3%

Table 3-6 represents a high level distribution of flight arrival time prediction errors. For active flights, there is no significant difference in prediction accuracy between April and June data: 89.7 to 92.7 percent of flights are on time, and practically all prediction errors are within ±1 hour range. For proposed flights, however, there is an obvious difference between April and June data. In April, there is a higher percentage on time predictions than in June: close to 60% in April vs. 53% in June. Additionally, in April, there is a lower percentage of late predictions than in June: approximately 30% in April vs. approximately 40% in June for late predictions. The difference between April and June predictions can be explained by the impact of traffic management initiatives (TMIs). A review of TMIs in April and June revealed a high number of ground stops on one of the June study dates (June 8th), which included several ground stops at the studied airports: 3 in ATL, 1 at DFW, MIA, ORD and SFO.

The distributions of prediction errors are presented in Figure 3-8 through Figure 3-11. Figure 3-8 and Figure 3-10 show the distributions of errors in airport arrival time predictions from April and June data, respectively. Figures 3-9 and 3-11 present the corresponding cumulative probability distributions. Each figure contains separate distribution curves for active and proposed flights and for two ranges of look-ahead prediction times: LAT = 1 – 2 hours and LAT = 2 – 3 hours.

These figures show that

- The prediction accuracy of airport arrival times for active flights is significantly better than for proposed flights that are still on the ground: for active flights, the prediction errors are within a ± 15-minute range, whereas for the proposed flights, the prediction errors are within a range from -45 min to 25 min.
- Within each case of active and proposed flights, the prediction accuracy is nearly the same for LAT = 1 – 2 hours and LAT = 2 – 3 hours.
- For active flights, the distributions of prediction errors are nearly symmetric with zero average error.

---

[2] "On-time" is considered to be within 15 minutes (early or late) of the predicted time.

- For the flights still on the ground, the distributions of prediction errors are biased (with median error of approximately -7 minutes) and asymmetric with heavier left-hand tails that reflect the tendency for proposed flights to arrive, on average, later than predicted.

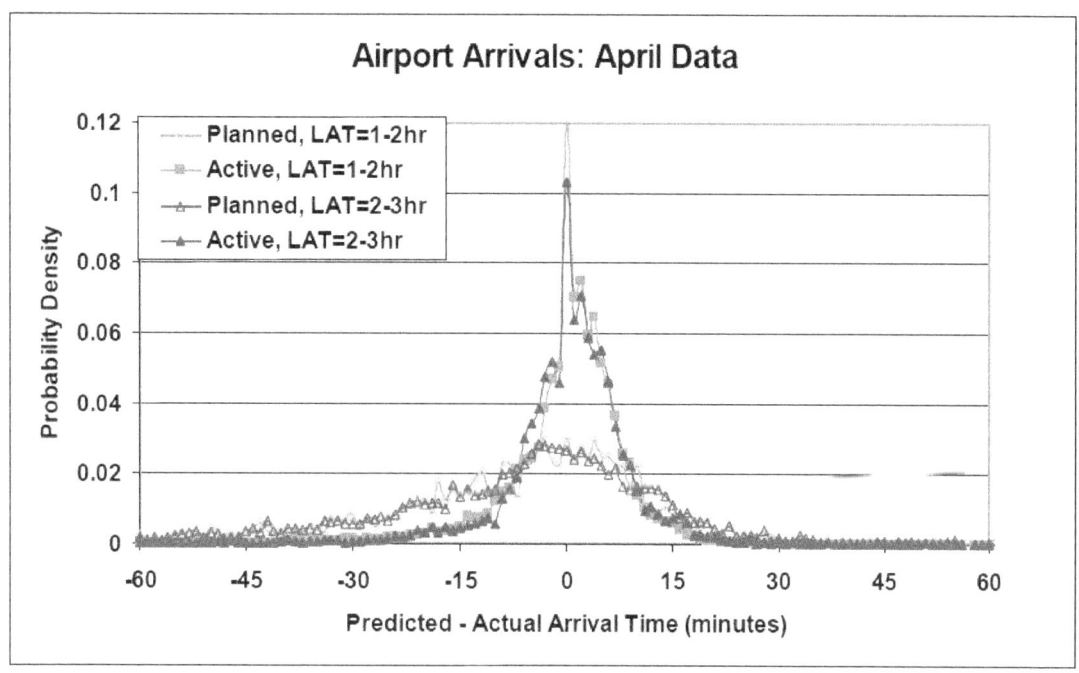

Figure 3-7 Distribution of Errors in Flight Arrival Time Predictions

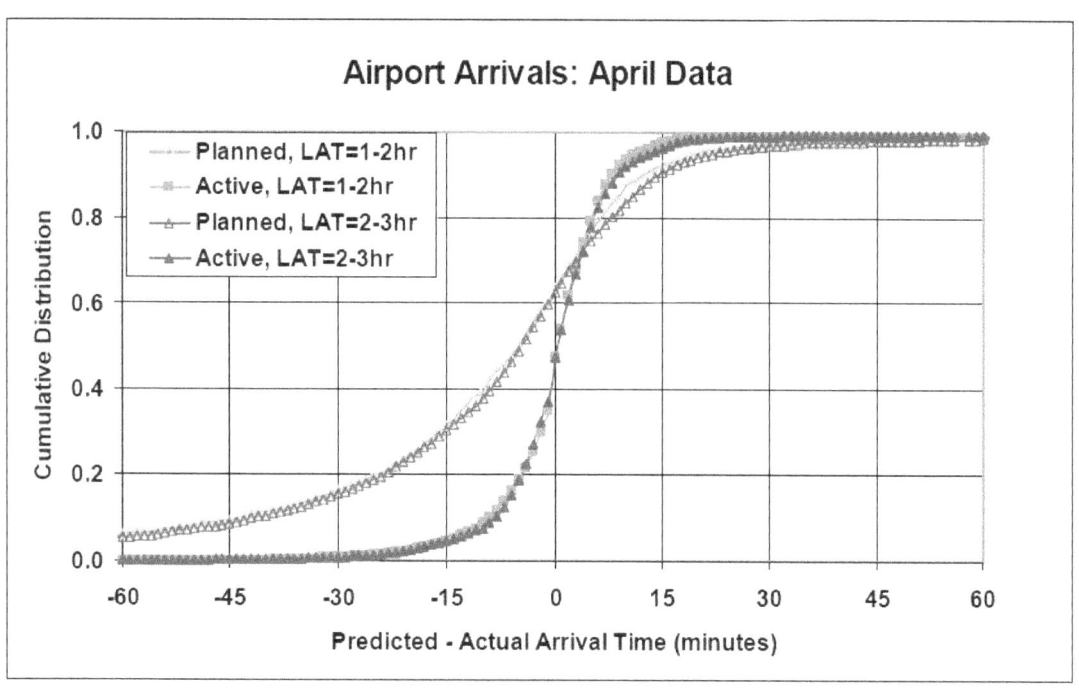

Figure 3-8 Cumulative Distribution of Errors in Flight Arrival Time Predictions

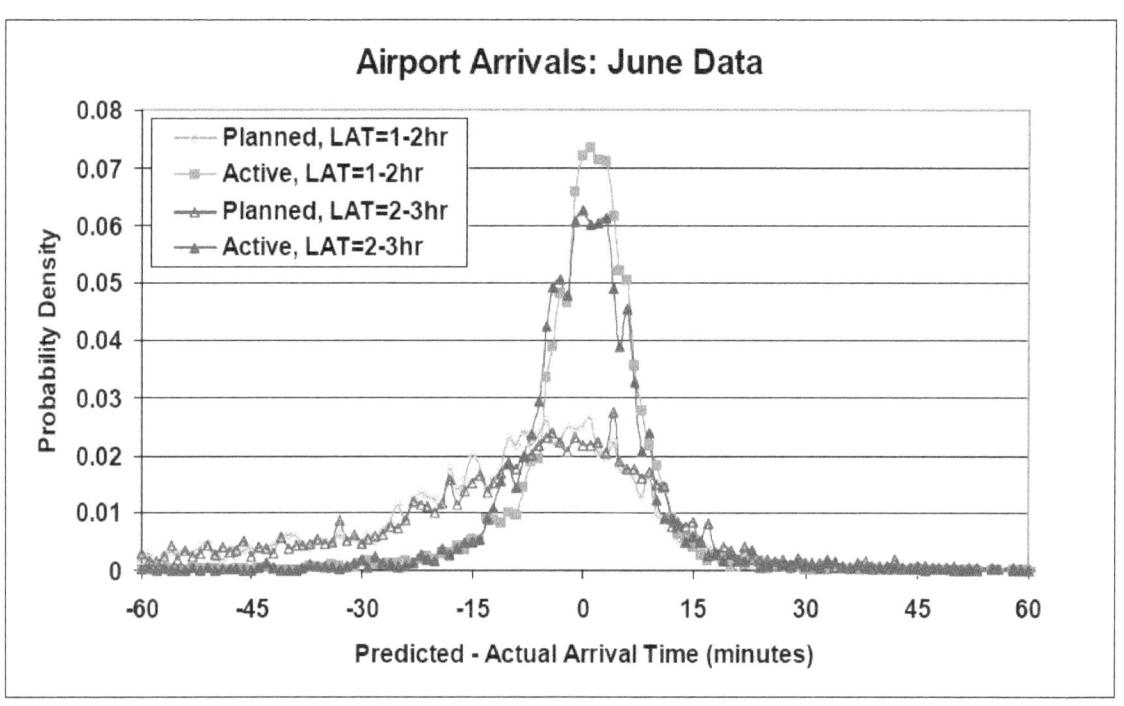

Figure 3-9 Distribution of Errors in Flight Arrival Time Predictions

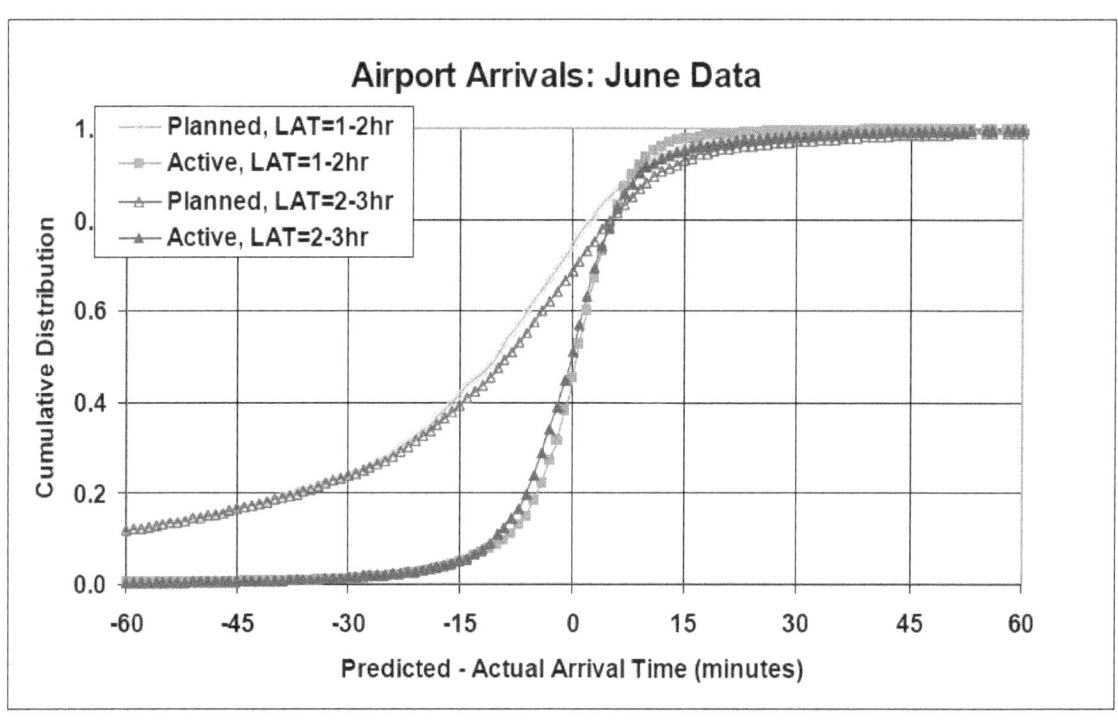

Figure 3-10 Cumulative Distribution of Errors in Flight Arrival Time Predictions

Figure 3-11 shows the cumulative probability distribution of prediction errors obtained from the combined historical data set that included the data for both active and proposed flights with look-ahead time prediction from 1 to 2 hours. The distribution is biased (with median error of approximately -3 minutes) and asymmetric with a heavier left-hand tail, toward early arrivals.

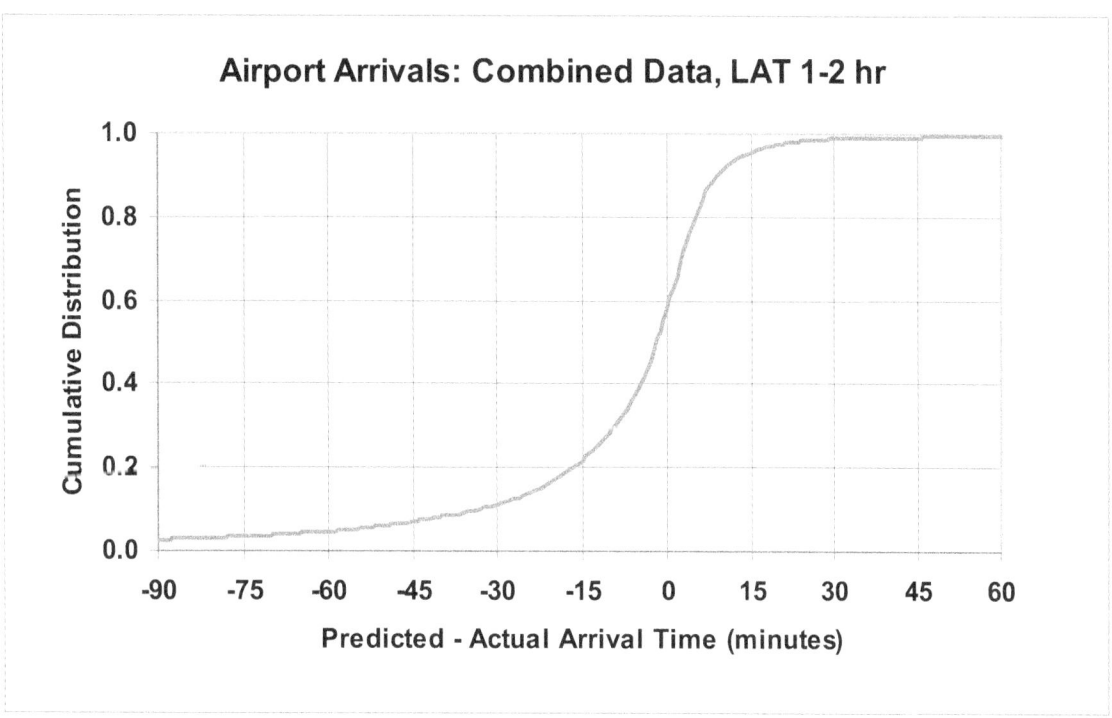

Figure 3-11 Cumulative Distribution of Errors in Flight Arrival Time Predictions

Table 3-7 through Table 3-10 present the results of statistical analysis of accuracy of flight arrival time predictions at each of nine airports considered in this study. Here again, there are significant differences in airport flight arrival time prediction accuracy between active and proposed flights. The differences are not only in absolute values of prediction errors but also in their signs: flights on the ground tend to arrive later than predicted, but the airborne flights tend to arrive earlier.

The differences in average error predictions vary by airport. In April, when there were many fewer TMIs than in June (see page 14), the error values at different airports vary within relatively narrow ranges for proposed and active flights, respectively. For example, in April, the prediction errors with LAT = 1 – 2 hours for the proposed flights vary from -13.9 to -10.6 minutes at seven of nine airports (except LAX and MIA). For the active flights with LAT = 1 – 2 hours, the average prediction errors vary from nearly zero to 3.9 minutes at seven of nine airports (with the exception of 7.4 minute early arrivals at MIA and 7.5 minute late arrivals at ORD). A relatively wider range of average prediction errors among airports occurred for LAT = 2 – 3 hours.

There is not much difference between LAT = 1 -2 hour and LAT = 2 – 3 hour predictions in each case of flights' status. However, there is a large difference in accuracy of prediction for flights still on the ground in April and June, when the June predictions are significantly less accurate than for April. The high number of ground stops on June 8th was noted earlier.

There is a large difference in prediction accuracy for active flights and for proposed flights. The differences in average prediction errors between airports vary from 3.6 to 16.7 minutes in April and from 5.8 to 33.4 minutes in June.

Table 3-7 April Data, Flights on the Ground, (Predicted-Actual) Arrival Time (minutes)

LAT	Dest. Airport	Number of Observations	Average Error (min)	Standard Deviation	Skewness
1 – 2 hr	ATL	3336	-11.0	23.1	-1.4
	BOS	1280	-12.9	23.5	-1.4
	DFW	1776	-10.6	20.6	-1.2
	LAX	1032	-5.9	16.2	-2.1
	MCI	710	-11.8	18.8	-1.3
	MIA	757	-5.2	28.2	-0.9
	ORD	1263	-13.9	43.6	0.9
	SFO	1132	-11.5	24.5	-1.3
	STL	1019	-12.5	22.1	-1.5
2 – 3 hr	ATL	2023	-1.7	19.4	-1.6
	BOS	880	-10.3	23.8	-0.9
	DFW	2239	-12.2	21.8	-1.3
	LAX	882	-7.2	20.6	-2.0
	MCI	517	-13.7	19.7	-0.8
	MIA	1213	-6.6	27.4	-0.7
	ORD	1227	-10.1	31.6	0.4
	SFO	507	-19.1	29.9	-0.9
	STL	779	-13.9	23.3	-1.4

Table 3-8 April Data, Flights in the Air, (Predicted-Actual) Arrival Time (minutes)

LAT	Dest. Airport	Number of Observations	Average Error (min)	Standard Deviation	Skewness
1 – 2 hr	ATL	1832	2.2	7.2	-2.3
	BOS	659	0.0	6.8	-1.9
	DFW	1300	3.9	6.5	2.0
	LAX	617	0.0	5.0	-1.5
	MCI	286	1.4	5.6	-2.5
	MIA	541	7.4	7.7	-1.0
	ORD	1178	-7.5	11.1	-4.1
	SFO	515	-0.4	8.3	-1.7
	STL	480	1.7	5.0	0.9
2 – 3 hr	ATL	916	4.2	6.9	-1.6
	BOS	373	0.0	8.4	-1.5
	DFW	1397	1.9	7.6	6.3
	LAX	694	-0.8	4.9	-1.9
	MCI	262	2.0	3.9	-0.4
	MIA	694	7.4	12.0	5.0
	ORD	826	-6.5	11.5	0.1
	SFO	260	-2.4	8.6	-1.0
	STL	254	2.0	8.2	1.2

Table 3-9 June Data, Flights on the Ground, (Predicted-Actual) Arrival Time (minutes)

LAT	Dest. Airport	Number of Observations	Average Error (min)	Standard Deviation	Skewness
1 – 2 hr	ATL	1721	-32.0	42.1	-1.2
	BOS	727	-22.2	36.1	-1.6
	DFW	969	-15.0	30.2	-2.9
	LAX	753	-9.3	15.8	-1.9
	MCI	352	-18.4	24.1	-1.5
	MIA	575	-19.0	35.4	-1.0
	ORD	977	-22.3	40.1	-1.1
	SFO	515	-12.3	23.5	-2.3
	STL	470	-16.1	26.5	-1.6

LAT	Dest. Airport	Number of Observations	Average Error (min)	Standard Deviation	Skewness
2 – 3 hr	ATL	2816	-27.4	41.5	-1.2
	BOS	1057	-21.9	38.1	-1.4
	DFW	2116	-15.4	29.7	-2.4
	LAX	1216	-6.5	18.0	-1.9
	MCI	625	-21.6	30.9	-1.8
	MIA	895	-21.3	36.0	-0.6
	ORD	1963	-19.6	43.7	-0.5
	SFO	780	-8.5	20.4	-2.5
	STL	739	-15.2	26.5	-1.6

Table 3-10 June Data, Flights in the Air, (Predicted-Actual) Arrival Time (minutes)

LAT	Dest. Airport	Number of Observations	Average Error (min)	Standard Deviation	Skewness
1 – 2 hr	ATL	1822	1.4	10.1	-2.2
	BOS	526	1.7	5.1	-0.1
	DFW	1498	1.7	5.5	1.0
	LAX	1006	-0.9	7.0	-9.4
	MCI	262	1.7	7.9	0.2
	MIA	624	-0.5	14.4	-1.6
	ORD	1449	-3.9	11.0	-3.2
	SFO	608	-0.4	5.5	0.4
	STL	315	5.4	8.4	1.1
2 – 3 hr	ATL	689	2.6	13.2	1.1
	BOS	320	1.9	9.2	-3.0
	DFW	735	2.3	8.4	1.2
	LAX	921	-2.0	11.0	-9.4
	MCI	113	4.7	12.6	1.3
	MIA	319	0.1	22.3	1.8
	ORD	891	-3.6	11.2	-0.6
	SFO	489	-2.7	7.4	0.9
	STL	121	11.7	14.1	0.9

4. Probability that a Flight will be in a Sector

Probabilistic traffic flow management (TFM) requires calculation of the probability that a flight will be in a sector at a given time. This probability involves two factors:

- The accuracy of the predicted sector entry time (discussed earlier)
- The accuracy of the predicted time-in-sector.

Time-in-sector is, of course, a derived measure based on the difference between exit and entry times for a particular flight. The data sets that were used earlier to support the analysis of sector entry also have information on sector exit times. These data sets were used to first, analyze time-in-sector, and then to calculate the probability that a flight will be in a sector given a predicted entry time.

4.1 Time in Sector

Sectors vary in size and in the manner flights traverse them. As a result, there are significant differences among the sectors in time-in-sector for a flight. Figure 4-1 shows the cumulative distributions of time-in-sector for the 9 sectors that were studied. This study used the April data. This figure represents, for each sector, the proportion of flights that are in the sector for less than the number of minutes indicated. For example, in ZMP20, 20% of the flights were in the sector for less than 15 minutes, while 70% of the flights were in the sector for less than 30 minutes. Other sectors had typically shorter in-sector times than ZMP20. For example, 70% of flights in ZBW46 were in the sector for less than 7 minutes.

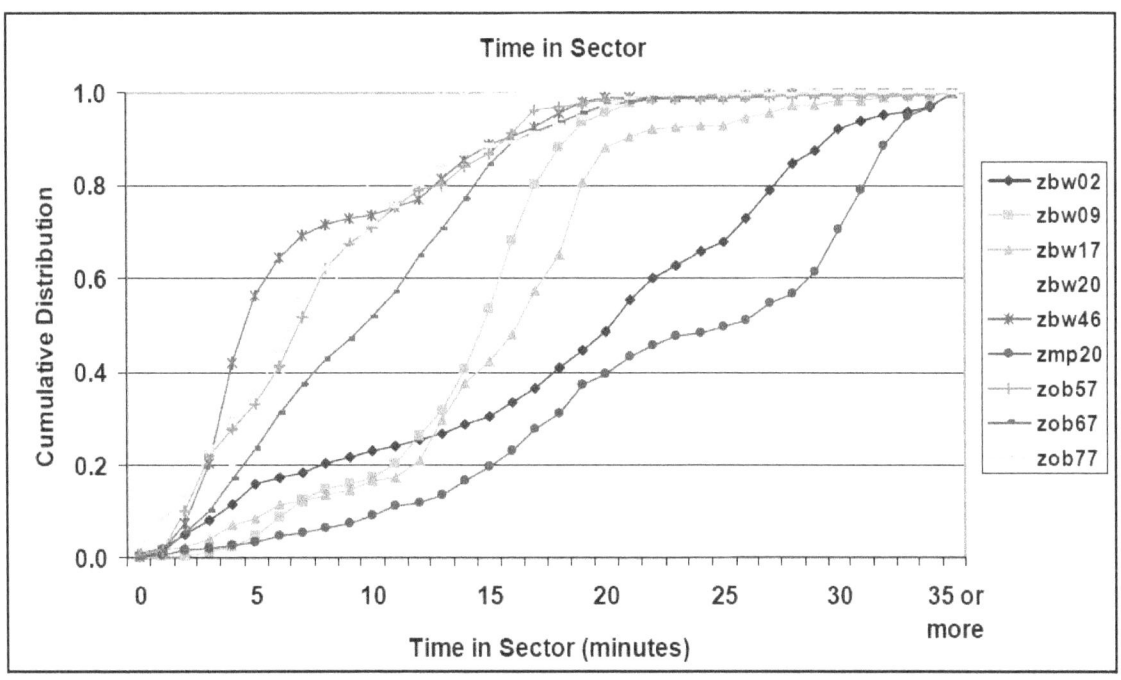

Figure 4-1 Cumulative Distribution of Time in Sector

Even though there is a wide distribution of actual time-in-sector, the prediction error for time-in-sector was smaller. Typically, the standard deviation for (Predicted − Actual) in-sector time would be 4 or 5 minutes, far smaller than the standard deviation of sector entry time (Table 4-1 and Table 4-2).

Furthermore, there was not much difference with respect to look-ahead time, or with respect to whether the flight was on the ground (proposed) or in the air (active). This is not surprising, as ETMS modeling of time-in-sector is unaffected by the look-ahead time or flight status.

Table 4-1 April Data, Flights on the Ground, (Predicted – Actual) In-Sector Time

Look Ahead Time	Sector	Number of Observations	Average Error (min)	Standard Deviation
1 – 2 hours	ZBW02	778	-0.03	3.88
	ZBW09	1293	-1.50	5.00
	ZBW17	777	-0.18	4.13
	ZBW20	2838	-0.16	1.91
	ZBW46	1567	-2.91	5.05
	ZMP20	875	-0.02	7.41
	ZOB57	1518	-0.69	5.98
	ZOB67	2149	-1.11	4.52
	ZOB77	1404	0.76	4.17
1 – 2 hr, combined		**13199**	**-0.72**	**4.70**
2 – 3 hours	ZBW02	691	-0.23	4.06
	ZBW09	1075	-1.93	4.81
	ZBW17	791	-0.27	3.87
	ZBW20	2461	-0.09	2.00
	ZBW46	1305	-2.83	4.98
	ZMP20	1137	-0.09	6.53
	ZOB57	1172	-0.58	7.41
	ZOB67	1806	-0.81	4.48
	ZOB77	1235	1.04	4.45
2 – 3 hr, combined		**11673**	**-0.63**	**4.87**
All data combined		**24872**	**-0.68**	**4.78**

Table 4-2 April Data, Flights in the Air, (Predicted – Actual) In-Sector Time

Look Ahead Time	Sector	Number of Observations	Average Error (min)	Standard Deviation
1 – 2 hours	ZBW02	853	-1.30	5.33
	ZBW09	192	-4.16	4.93
	ZBW17	226	-0.01	6.54
	ZBW20	641	0.35	1.94
	ZBW46	205	-2.41	4.91

Look Ahead Time	Sector	Number of Observations	Average Error (min)	Standard Deviation
	ZMP20	811	-0.35	6.02
	ZOB57	76	-1.54	5.42
	ZOB67	236	-1.40	4.07
	ZOB77	291	0.63	2.95
1 – 2 hr, combined		3531	-0.77	5.01
2 – 3 hours	ZBW02	822	-1.35	4.95
	ZBW09	87	-4.10	4.36
	ZBW17	103	0.07	7.89
	ZBW20	489	0.07	1.94
	ZBW46	141	-1.74	3.94
	ZMP20	240	-0.62	8.54
	ZOB57	14	1.14	9.22
	ZOB67	135	-2.00	3.83
	ZOB77	66	-0.33	2.67
2 – 3 hr, combined		2097	-1.00	5.12
All data combined		5628	-0.86	5.05

4.2 Probability that a Flight will be in a Sector

ETMS predicts that a flight will be in a sector during a particular time interval. The accuracy of this prediction is a function of errors in both the predicted entry and exit times. Using empirical data on entry and exit predictions, we calculate the probabilities that a flight will be in a sector at a given time. Flights are grouped by predicted time in-sector. Predictions for active and proposed flights are presented separately, because there is much more variability in the predictions of sector entry for proposed flights. The following groupings by time-in-sector are used, both for all sectors combined and for selected individual sectors.

- 30 – 34 minutes (primarily ZMP20)
- 15 – 19 minutes
- 10 – 14 minutes
- 5 – 9 minutes
- 0 – 4 minutes.

Figures 4-2 through 4-6 show the probabilities, beginning on page 24.

Figure 4-2 shows the fraction of flights in the sector as a function of predicted sector entry times for those flights that had a predicted in-sector time in the 30-34 minute range. These were primarily flights traversing ZMP20. If predictions were perfect, we would expect this curve to have the following shape:

- Before time 0, there are no flights in the sector, because they have not reached the sector yet
- At time 0, flights enter the sector, and the curve jumps to 100%

- Somewhere between times 30 and 34, flights leave the sector, so the curve falls from 100% to 0% between times 30 and 34.

The curves for active flights are close to this ideal, with, for example, some 99% of flights that were predicted to be in the sector at time 15 actually being in the sector. For proposed flights, the curve falls short of the ideal, with less than 70% of those flights predicted at time 15 actually being in the sector.

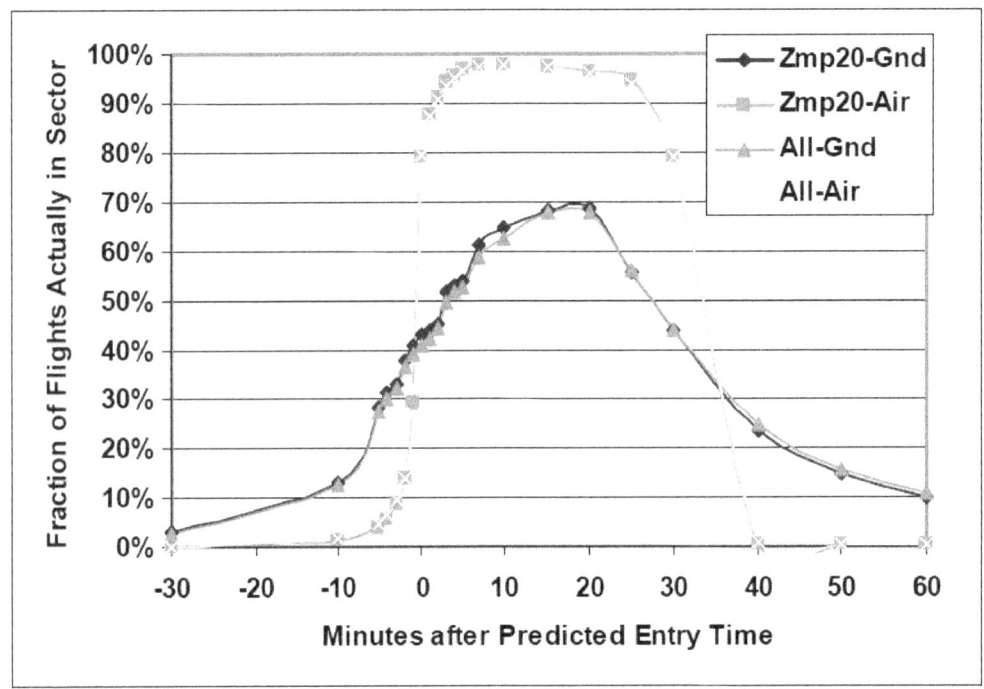

Figure 4-2 Predicted In-Sector Time: 30 – 34 minutes

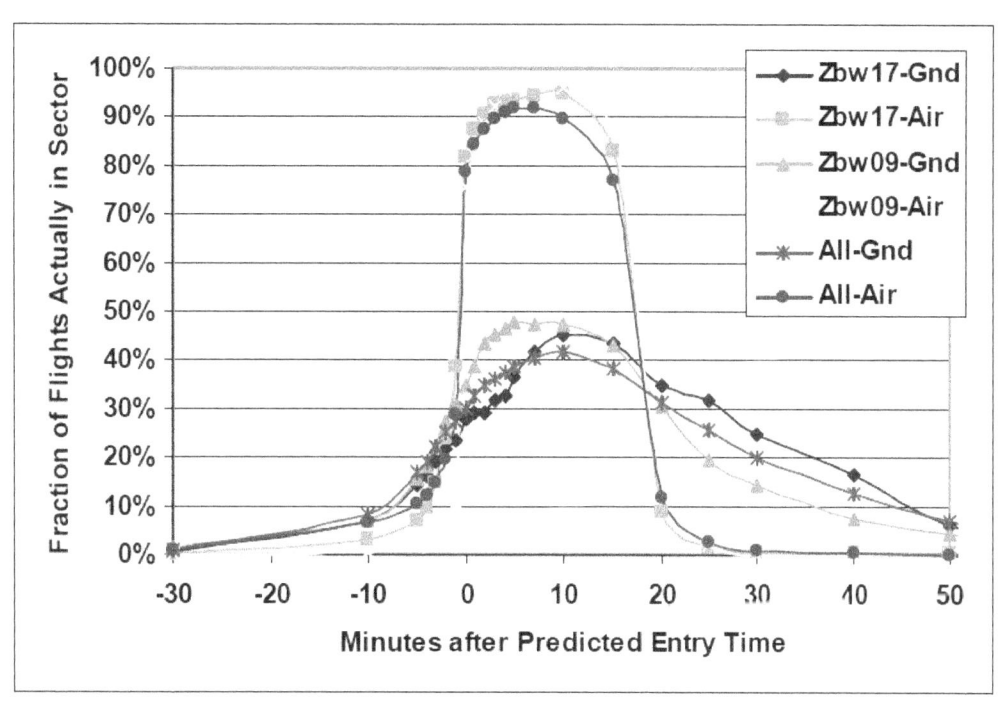

Figure 4-3 Predicted In-Sector Time: 15 – 19 minutes

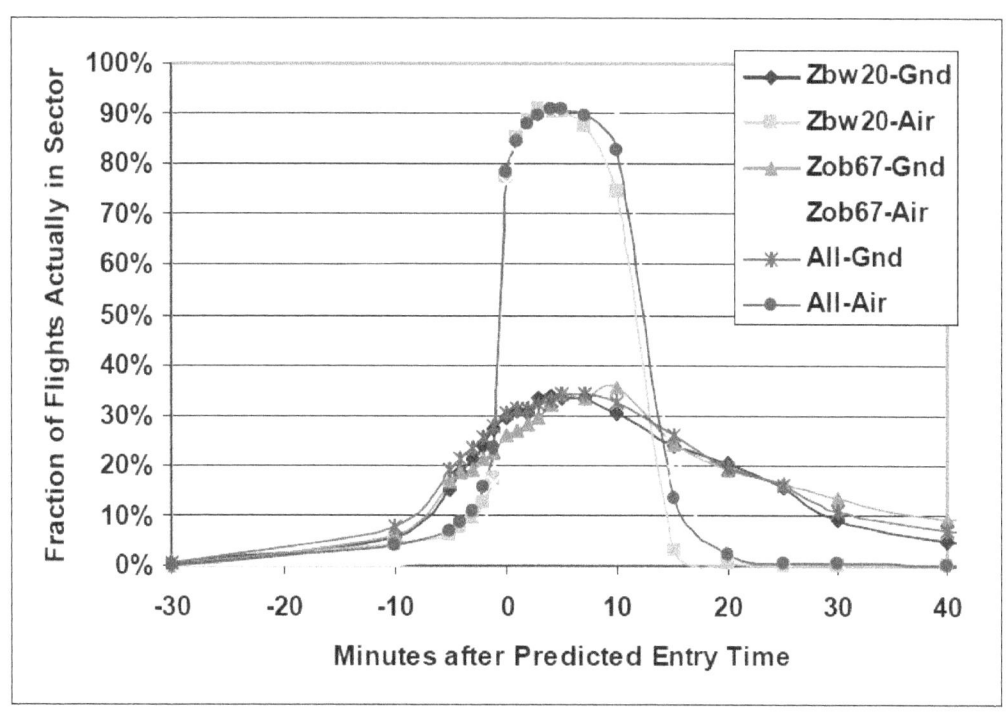

Figure 4-4 Predicted In-Sector Time: 10 – 14 minutes

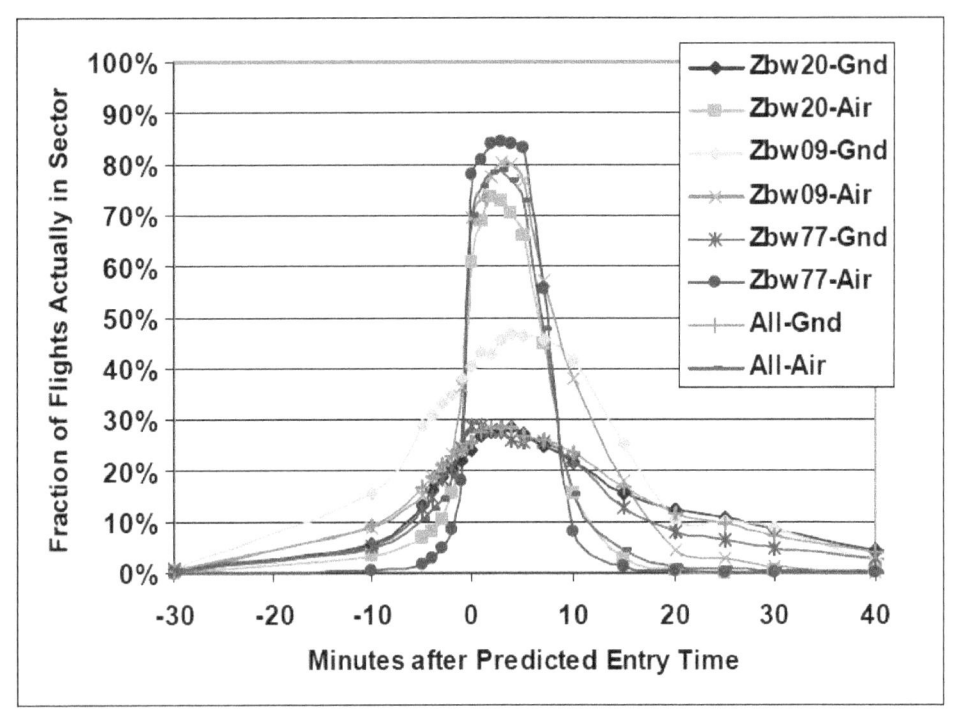

Figure 4-5 Predicted In-Sector Time: 5 – 9 minutes

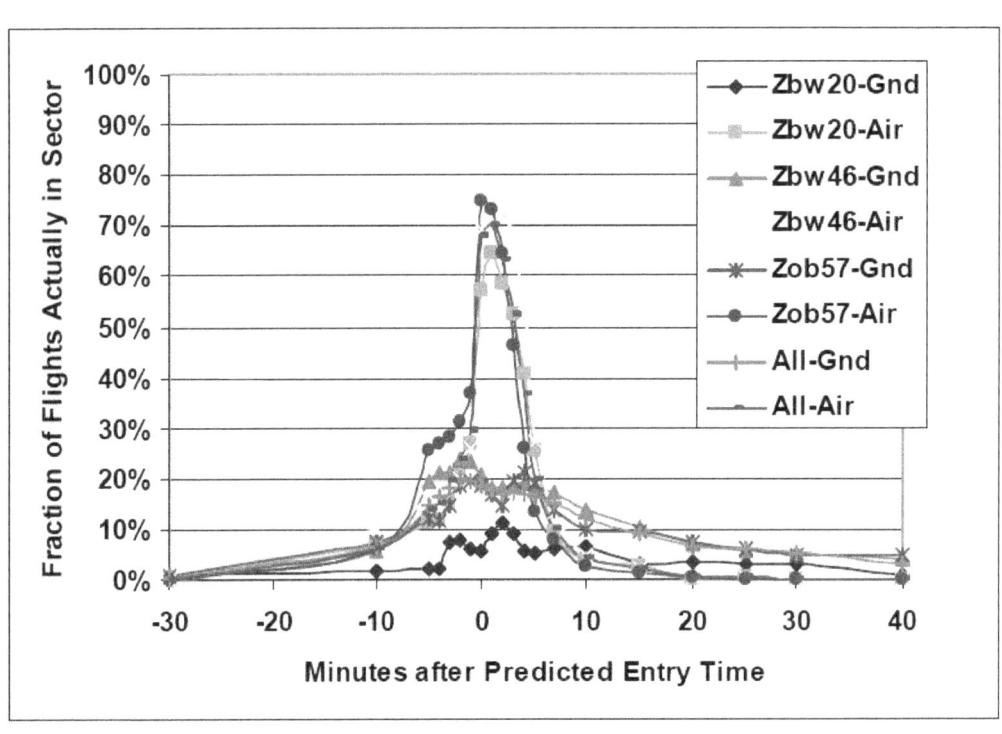

Figure 4-6 Predicted In-Sector Time: 0 – 4 minutes

5. Probabilistic Count Predictions

5.1 Introduction

Previous sections analyzed uncertainty in predicting events for individual flights, such as distributions of errors in flight arrival time at airports or in the times of crossing sector boundaries. Now, these results will be used for probabilistic prediction of traffic demand counts.

The main difference between deterministic and probabilistic predictions is as follows: Deterministic demand counts are completely based on aggregating the flights in the time interval (e.g., 15-minute interval) for which deterministically Estimated Time of Arrivals (ETA) fall within the interval. Probabilistic predictions, however, take into account the probability distributions of random errors in ETA predictions. These distributions are then used for estimating probabilities for individual flights to arrive in the time interval, which, in turn, are used for computing probability distribution of predicted demand counts for the interval. A comparative analysis of deterministic and probabilistic demand predictions was presented in [3]. Stochastic simulation results, presented in [3], showed that probabilistic predictions out-performed the deterministic ones in terms of their accuracy. Along with the stochastic simulation results, the author of the paper [3] presented an analytical method to calculate probabilistic predictions of demand counts.

This analytical method (with some simplifications) is used in this section for probabilistic demand counts prediction. An alternative method based on approximation of binomial distribution by a normal (Gaussian) distribution is also presented in this section.

The end objectives of this study are to (a) understand the relationship between the accuracy of flight-by-flight predictions and those of aggregate count predictions, and (b) use this understanding to develop probabilistic demand count predictions.

The steps taken in this section are as follows:

1. Translate a flight's time predictions and associated prediction errors into the probabilities for the flight to be in particular 15-minute time intervals.
2. Show how these probabilities can be combined to develop probabilistic count predictions for an interval (using analytical approach described in [3]).
3. Simplify this approach by considering the normal approximation to the binomial distribution. Derive means and standard deviations of demand counts.
4. Show how the derived probabilistic count predictions relate to the empirically observed count predictions.

5.2 Probability that a Flight will Arrive During a Specified Interval

Consider a flight that is deterministically forecast to arrive at a NAS element (airport or sector boundary) at time $\mathbf{x}$, i.e., $\mathbf{x}$ is equal to a flight's ETA. What is the probability that the flight can actually arrive during the interval $[\mathbf{a}, \mathbf{b})$ [3]? We denote this conditional probability as Prob(Actual in $[\mathbf{a}, \mathbf{b})$ | fc at time $\mathbf{x}$), where "fc" means forecast.

Considering the errors in flight's arrival time prediction (presented in Section 3 of this report), the flight can actually arrive during the interval in question if the prediction error is between $\mathbf{x} - \mathbf{b}$ and $\mathbf{x} - \mathbf{a}$. Using

---

[3] The left bracket "[" indicates that the interval includes $\mathbf{a}$, while the right parentheses ")" indicates that the interval goes up to but does not include $\mathbf{b}$.

the cumulative probability distribution function (CDF) of prediction error presented in Figure 3-11, this probability can be determined as follows:

$$\text{Prob}(\text{Actual in } [a,b) \mid \text{fc at time } x) = \text{CDF}(x - a) - \text{CDF}(x - b).$$

ETMS determines air traffic demand by aggregating individual flights in 15-minute intervals. Therefore, the intervals of interest for actual arrivals of individual flights will be the 15-minute intervals $[a, a + 15)$ with $b = a + 15$, where a is a Zulu time that determines the start of a 15-minute interval (e.g., 1200, 1215, 1230, etc.). Hence, if a flight is predicted to arrive at time x, the probability for the flight to actually arrive in a 15-minute interval $[a, a + 15)$ is:

$$\text{Prob}(\text{Actual in } [a, a + 15) \mid \text{fc at time } x) = \text{CDF}(x - a) - \text{CDF}(x - a - 15).$$

For example, if $x = 1140$, $a = 1200$ and $b = 1215$, the flight can actually arrive in the [1200, 1215) interval if it is 20 to 35 minutes late, that is, if its "Predicted – Actual" time is between -20 and -35 minutes. This probability is

$$\text{Prob}(\text{Actual in } [1200, 1215) \mid \text{fc in time } 1140) = \text{CDF}(-20) - \text{CDF}(-35).$$

In a second example, $x = 1205$, $a = 1200$ and $b = 1215$. Here, the flight can actually arrive in the [1200, 1215) interval if it is between 5 minutes early and 10 minutes late; that is, if its "Predicted – Actual" time is between 5 and -10 minutes. This probability is

$$\text{Prob}(\text{Actual in } [1200, 1215) \mid \text{fc in time } 1205) = \text{CDF}(5) - \text{CDF}(-10).$$

Figure 5-1 illustrates how the probabilities relate to the CDF for these two examples. In the first example, the probability is equal to (CDF(-20) – CDF(-35)) and is approximately 8% (the short vertical red line). In the second example, the probability is equal to (CDF(5) – CDF(-10)) and is approximately 50% (the longer vertical green line).

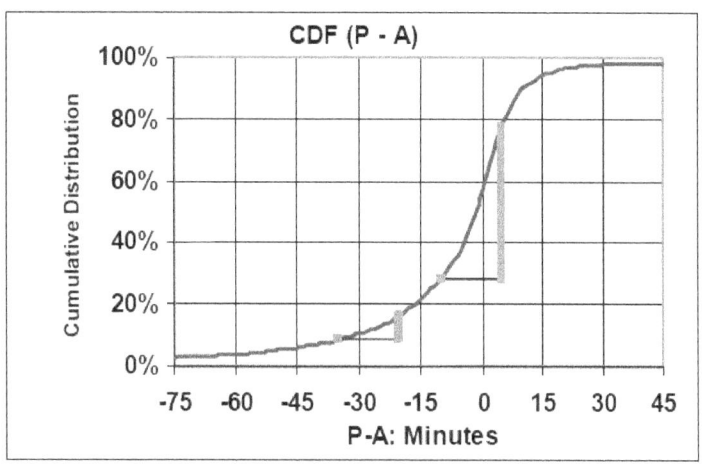

Figure 5-1 Predicted-Actual (P-A) Time and Flight Arrival Probability

The red curve in Figure 5-2 illustrates the distribution of probabilities for individual flights to arrive at [1200, 1215) interval as a function of predicted arrival time **x**. This curve, which is simply a plot of all of the possible vertical lines from Figure 5-1, makes it possible to determine the probability for a flight with specific predicted arrival time **x** to actually arrive in the [1200, 1215) interval. These probabilities depend on how far the predicted arrival time is from the 15-minute interval of interest for actual arrivals: the farther the time **x** from the 15-minute interval for actual arrivals the smaller the probability is for the flight to actually arrive during the interval. Using the two previous examples, if **x** is 1140, the probability is 8%, while if **x** is 1205, the probability is 50%.

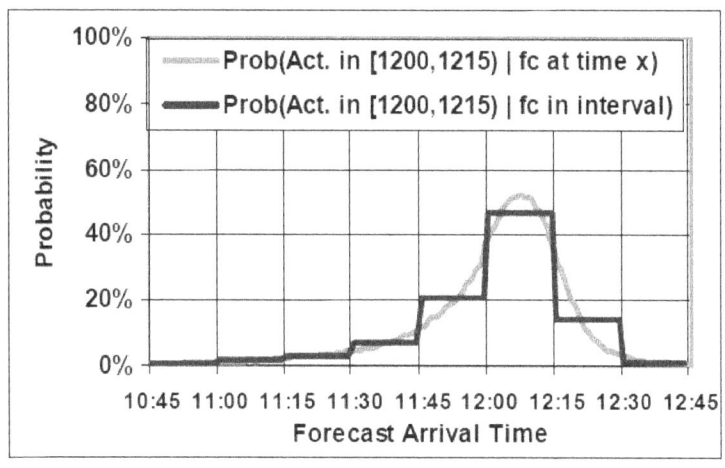

Figure 5-2 Relationship between Flight Arrival Probability and Estimated Arrival Time

It is worth noticing that the major reason for determining probabilities for individual flights to arrive in a specific 15-minute interval is to later use those probabilities to aggregate the flights and then develop a probabilistic prediction of aggregate 15-minute demand counts. Theoretically, it is possible to calculate the probabilistic distribution of aggregate traffic demand by using different probabilities for individual flights (see Section 5.3). However, practically it will require cumbersome computations if a large number of flights are involved. Therefore, to simplify the computational procedure, we will work with the probabilities that are the same for each flight in a group of flights predicted to arrive within a particular 15-minute interval, not at the particular time **x**. As a result, the probabilities for individual flights will be different only when they are predicted to arrive in different 15-minute intervals. Those probabilities can be obtained from the distribution of probabilities to actually arrive within a particular 15-minute interval as a function of time **x** (see the red curve in Figure 5-2) through the step-wise approximation of the curve so that the function remains constant within each 15-minute interval. The step-wise approximation is shown in Figure 5-2 by the blue line. The constant values for the approximation can be determined

- by averaging the red function at each 15-minute interval, or
- by assigning a constant probability in accordance with the allocation of predicted arrival times for individual flights within the interval.

The step-wise approximation makes it easier to deal with the number of flights that are forecast to arrive in the 15-minute interval [**y**, **y** + 15). The constant value that corresponds to this interval determines the probability for each flight predicted at this interval to actually arrive within the [**a**, **a** + 15) interval. This probability is denoted as

Prob(Actual in [**a**, **a** + 15) | fc in [**y**, **y** + 15)).

For example, Figure 5-2 gives the following probabilities for individual flights to actually arrive in [1200, 1215) interval if they are forecasted to arrive at various 15-minute intervals:

Prob(Actual in [1200, 1215) | fc in [1130, 1145)) = 0.07

Prob(Actual in [1200, 1215) | fc in [1145, 1200)) = 0.21

Prob(Actual in [1200, 1215) | fc in [1200, 1215)) = 0.47

Prob(Actual in [1200, 1215) | fc in [1215, 1230)) = 0.14

Note that if the flight is forecast to arrive in the interval [1200, 1215), it only has slightly less than 50% probability of actually arriving in that interval.

The above numbers, as well as the curves in Figure 5-2, are obtained for the accuracy of predictions for individual flights characterized by cumulative probability distribution of prediction errors in ETMS and shown in Figure 3-11.

The next question is how the improvement in accuracy of flight arrival time prediction affects the predictability for individual flights to actually arrive at a particular 15-minute interval.

Figure 5-3 shows what happens if flights' arrival time predictions are twice as accurate, i.e., the standard deviation of the arrival time prediction error is a one half of the one that corresponds to the CDF in Figure 5-1.

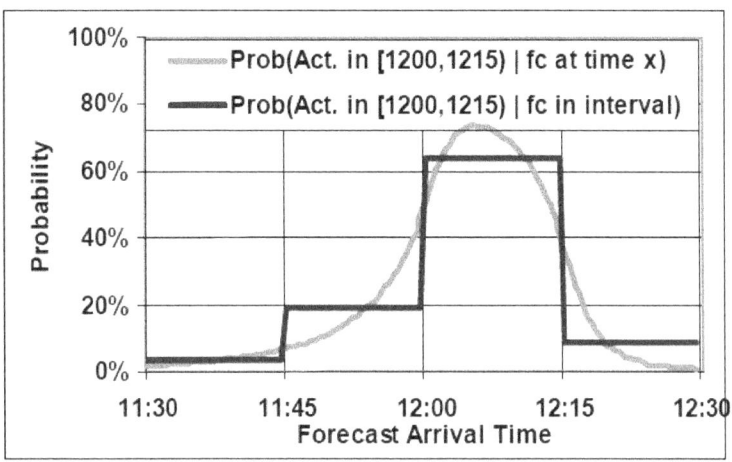

Figure 5-3 Flight Arrival Probability with More Accurate Arrival Time Predictions

Both Figures 5-2 and 5-3 reflect the relationship between the accuracy of a prediction for an arrival time, and the accuracy of a prediction for arrival within a specific time interval. Figure 5-3 shows the impact of more accurate predictions, for example, that if a flight is forecast to arrive in the interval [1200, 1215), it now has a 64% probability of actually arriving in that interval, versus the 47% probability shown in Figure 5-2.

5.3 Combining Probabilistic Count Predictions

In the previous section, we examined the probability for an individual flight to arrive within a given interval, given its forecast arrival time. In this sub-section, we focus on combining the probabilities for those individual flights, to develop a probabilistic count prediction.

The idea of Meyn's [3] method for combining probabilistic count predictions is as follows. Suppose that for certain set of flights (e.g., N flights), the probabilities for each individual flight to arrive at a NAS element (such as an airport or sector boundary) during a time interval are known, so that the probability for flight i to arrive during the time interval is equal to p_i (i = 1, 2, ..., N). For example, if flight i is predicted to arrive at 1205, and our interval of interest is 1200 – 1215, then the previous section (5.2) tells us that p_i = 0.50. Given the probabilities p_i for individual flights, the probability distribution of demand counts will then be calculated, which will determine probability of each possible demand forecast count from 0 to N out of N flights. As a result, a set of probabilities $P_N[k]$ that k of N flights are predicted to arrive at an airport or enter a sector determine the probabilistic demand forecast (k = 0, 1, 2, ..., N). Assuming that the events for flights to arrive at a NAS element are independent, the following computationally economical recursive procedure for calculating probabilities $P_N[k]$ was presented in [3]:

$P_0[0] = 1$

For i = 1 to N:

$P_i[0] = (1 - p_i) P_{i-1}[0]$

$P_i[i] = p_i P_{i-1}[i-1]$

For k = 1 to i-1:

$P_i[k] = p_i P_{i-1}[k-1] + (1 - p_i) P_{i-1}[k]$.

This procedure covers a general case when the probabilities p_i (i = 1, 2, ..., N) are different. When all flights have the same probability $p_i = p$ (i = 1, 2, ..., N) to arrive at a NAS element, the above recursive procedure reduces to binomial distribution:

$$P_N[k] = \frac{N!}{k!(N-k)!} p^k (1-p)^{N-k}, \quad k = 1, 2, 3, ..., N \tag{1}$$

where $k! = 1 * 2 * 3 * ... * k$.

The above probabilities can be applied to both sector entry counts and airport arrivals. However, traffic demand counts for sectors is currently measured by the number of flights in the sector for each minute, and the maximum one-minute count of fifteen one-minute predictions is taken as traffic demand count for a sector. This measure is under consideration for revision and probabilistic sector demand count prediction can be addressed in future research. In the remainder of this section, we focus on 15-minute traffic demand count predictions at arrival airports.

5.4 A Tractable Probabilistic Count Prediction

Consider the universe of aircraft that might actually arrive during the interval [**a**, **a** +15). The universe may include the flights that are predicted to arrive during [**a**, **a**+15), flights predicted to arrive during adjacent intervals, and other flights, such as pop-ups. There are several ways to calculate the probability distribution of the number of aircraft predicted to actually arrive during this interval.

- Consider each aircraft individually, and use Meyn's approach [3] to develop the probabilistic prediction.
- Group the aircraft by forecast arrival time buckets (see Section 5.2)
- Limit the number of arrival time buckets used
- Only consider those aircraft that are forecast to arrive during the interval [**a**, **a** +15).

If aircraft are grouped by forecast arrival time bucket, the probability distribution of the actual arrival counts at a specific time interval is the sum of independent random values with the probability distributions that are specific for each bucket (the probability distribution of number of flights with ETAs within the bucket that would arrive at the time interval of interest). The distribution of the sum of binomially distributed values is difficult to calculate. However, the calculations can be significantly simplified if the normal (Gaussian) approximation of binomial distribution is used. A binomial distribution with parameters n and p (e.g., n is the total number of flights in a set, and p is the probability of a flight to arrive at the time interval) can be approximated by a normal distribution, with mean np and variance np(1-p).

Now, suppose that there are J interval buckets around the 15-minute interval of interest [**a**, **a** + 15) (including this interval) with n_j flights within the jth bucket and each flight has a probability p_j (j = 1, 2, ..., J) to arrive during the 15-minute interval of interest. For each bucket, the random number of the flights from the bucket that could arrive at the [**a**, **a** + 15) interval has a binomial distribution that can be approximated by a normal distribution. Then the expected total number of aircraft E(**a**) that would arrive at the [**a**, **a** + 15) interval (starting at **a**) can be approximated by

$$E(\mathbf{a}) = \sum_{j=1}^{J} n_j p_j \tag{2}$$

Similarly, the variance $\sigma^2(\mathbf{a})$ of the number of aircraft is approximated by

$$\sigma^2(\mathbf{a}) = \sum_{j=1}^{J} n_j p_j (1-p_j), \qquad (3)$$

where σ(a) is a standard deviation of the number of aircraft, and it is equal to square root of the variance

$$\sigma(\mathbf{a}) = \sqrt{\sum_{j=1}^{J} n_j p_j (1-p_j)} \qquad (3')$$

In these equations, again:

n_j - number of flights with ETAs in the jth bucket

p_j – probability that a flight in the jth bucket will arrive in [a, a +15)

Return to the example from section 5-2 to estimate the expected number of aircraft that could arrive during the [1200, 1215) interval. The step-wise function from Figure 5-2 gives the approximate probabilities for the flights with ETAs in a specific 15-minute time bucket to arrive during the [1200, 1215) interval. Table 5-1 shows the buckets of forecast arrivals, the number of forecast aircraft at each bucket (deterministic forecast), and the probabilities from Figure 5-2. In Table 5-1, D(a) denotes a deterministically predicted arrival demand counts at the 15-minute bucket that starts at time a. Demand D(a) is equal to the number of the flights with ETAs in the bucket.

Table 5-1 Arrival Time Buckets and Probabilities from Figure 5-2

n: Number of forecast aircraft	Forecast arrival bucket	p_j: Probability that actual arrival is in [1200,1215)
D(1045)	1045 – 1100	0.01
D(1100)	1100 – 1115	0.02
D(1115)	1115 – 1130	0.03
D(1130)	1130 – 1145	0.07
D(1145)	1145 – 1200	0.21
D(1200)	1200 – 1215	0.47
D(1215)	1215 – 1230	0.14
D(1230)	1230 – 1245	0.01
D(1245)	1245 – 1300	0.00
D(other)	Not in 1045 - 1300	Small

D(1045) represents number of flights with ETAs in the first (j = 1) 15-minute bucket that starts at 1045; D(1100) represents number of flights with ETAs in the second (j = 2) 15-minute bucket that starts at 1100, etc.. Hence, according to above notation:

D(1045) = n_1, D(1100) = n_2, D(1115) = n_3, …, D(1245) = n_9.

The first line of this table represents aircraft forecast to arrive in the interval 1045 – 1100Z. These aircraft have a 1% probability of arriving in the interval 1200-1215. On the other hand, aircraft that are forecast to arrive in the interval 1200-1215 have a much higher probability (47%) of arriving in that interval. The last line represents those aircraft that are not forecast to arrive in any of the listed arrival time buckets.

One can simplify the presentation in Table 5-1 by reducing the number of buckets to J = 3, and neglecting the buckets with small probabilities for arriving in [1200,1215) interval, for example:

Table 5-2 Arrival Time with Three Buckets

n: Number of forecast aircraft	Forecast arrival bucket	p_j: Probability that actual arrival is in [1200,1215)
D(1145)	1145 – 1200	0.21
D(1200)	1200 – 1215	0.47
D(1215)	1215 – 1230	0.14
D(other)	Not in 1145 – 1230	Small

In the case of Table 5-2 and according to equation (2), the expected number of aircraft for the interval starting at 1200, E(1200), is then given by

$$E(1200) = 0.21\, D(1145) + 0.47\, D(1200) + 0.14\, D(1215) + pD(other), \tag{4}$$

where pD(other) is a to-be-determined adjustment, representing the last line in Table 5-2 with small probability p.

Equation (3) gives the following formula for the variance $\sigma^2(1200)$ of the aircraft count for the interval starting at 1200:

$\sigma^2(1200) =$

$$0.21(1- 0.21)\, D(1145) + 0.47(1- 0.47)\, D(1200) + 0.14(1- 0.14)\, D(1215) + p(1-p)D(other) \tag{5}$$

Equation (5) can be simplified by noting that in the last term, $p(1-p)D(other)$, p is assumed to be much smaller than 1 and, hence, $(1-p) \approx 1$, so this term can be approximated by pD(other). As a result, Equation 5 is transformed to

$$\sigma^2(1200) = 0.17\, D(1145) + 0.25\, D(1200) + 0.12 D(1215) + pD(other). \tag{6}$$

What is pD(other)? D(other) may represent a number of aircraft, each with a small probability **p** of actually arriving during the 1200,1215 interval, and pD(other) can be interpreted as a residual error of the model.

There is a variety of possible ways of dealing with the residual error in the model. One approach is to use the equations 4 and 6 with the shown coefficients and neglect the last term pD(other). This will create bias in the resulting model, since the pD(other) term is positive. A second approach is to use models 4 and 6 without the last term pD(other) but adjust the coefficients to try to compensate for the error, using the following approach:

Since we have observed that the ETMS predictions are not significantly biased, one might reasonably assume that the sum of the coefficients of the terms D(1145), D(1200) and D(1215) in Equation 4 should be close to 1. The adjusting multiplier β to those coefficients can be found from the following equation:

$$\beta\,(0.21 + 0.47 + 0.14) = 1,$$

from which β = 1.22.

Equation (4) then becomes

E(1200) = 1.22 (0.21 D(1145) + 0.47 D(1200) + 0.14 D(1215)) =

= 0.26 D(1145) + 0.57 D(1200) + 0.17 D(1215), (7)

and the residual error pD(other) can be found from the following equation

pD(other) = 0.22 (0.21 D(1145) + 0.47 D(1200) + 0.14 D(1215)) (8)

Equation (6) becomes

σ^2(1200) = 0.21 D(1145) + 0.35 D(1200) + 0.15 D(1215) (9)

What if one only uses the forecast aircraft in the [1200,1215) bucket, as ETMS does? Then the expected number of flights predicted to arrive within [1200,1215) interval is less than half of the deterministically predicted demand counts D(1200):

E(1200) = 0.47 D(1200)

The adjusting multiplier β for this coefficient can be found from the following equation:

$$0.47 \beta = 1,$$

from which β = 2.13, and

pD(other) = (β –1)(0.47 D(1200)) = 0.53 D(1200) (10)

The variance of predicted demand counts in this case, with J = 1, is equal to

σ^2(1200) = (0.47(1–0.47) + 0.53) D(1200) = 0.78 D(1200), (11)

and the standard deviation is equal to

σ (1200) = 0.88 $\sqrt{D(1200)}$ (12)

Note, that both the expected demand counts and standard deviation of probabilistically predicted demand counts depend on deterministically predicted demand counts at all 15-minute intervals used in the probabilistic model.

With the assumption that the probability distribution of demand counts is close to a normal distribution, the knowledge of an expected value and a variance of predicted demand counts make it possible to obtain a probabilistic prediction of the counts by using the table of percentiles for the normal distribution. If the expected value is equal to E and standard deviation is σ, then the probability that the predicted demand counts will be within the range of (E ± m σ), where m determines the desired range in the numbers of standard deviations, can be found from the table for the normal distribution. For example, if m=1, then the probability is equal to 0.68. If m = 2, the probability is equal to 0.95, i.e., with the probability of 0.95, demand counts will be within the range from (E - 2σ) to (E + 2σ).

If one wants to know what is the threshold value for demand counts so that the predicted demand cannot exceed it with the certain probability, then the threshold value can be found from the table as a corresponding percentile. For example, the demand threshold that cannot be exceeded with the probability more than 0.25 is the 75th percentile, and according to the table is equal to (E + 0.68 σ). The threshold that determines the lower bound for demand counts to be below the threshold with 0.4 probability (not to be below this value with 0.6 probability) is the 40th percentile, which is equal to (E – 0.26 σ). The expected value E itself represents the 50th percentile. It also worth noticing that, for symmetric, bell-shaped distributions, the expected value is the most likely value among other values of this distribution..

To illustrate how the expected value and the uncertainty range of probabilistic demand count predictions depend on the deterministic predictions, consider numerical examples with three cases of deterministic demand predictions for three consecutive 15-minute intervals (Table 5-3).

Table 5-3 Examples of Probabilistic Predictions

	Case 1	Case 2	Case 3
D(1145)	25	12	25
D(1200)	20	20	20
D(1215)	12	12	30
E(1200) from equation 7	19.9	16.6	23.0
σ^2(1200) from equation 9	14.3	11.4	16.9
Standard Deviation σ(1200)	3.8	3.4	4.1
2.3th percentile (E – 2σ)	12.3	9.8	14.8
25th percentile (E - 0.68σ)	17.3	14.3	20.2
75th percentile (E + 0.68σ)	22.5	18.9	25.8
97.7th percentile (E + 2σ)	27.4	23.3	31.2

In Case 1, D(1145) = 25; D(1200) = 20; D(1215) = 12.

According to equations (7) and (9), the expected value and variance of probabilistically predicted demand counts for the 15-minute interval that starts at 1200 are:

E(1200) = 0.26*25 + 0.57*20 + 0.17*12 = 6.5 + 11.4 + 2.0 = 19.9

σ^2(1200) = 0.22*25 + 0.35*20 + 0.15*12 = 5.5 + 7.0 + 1.8 = 14.3.

The standard deviation σ(1200) is equal to

σ (1200) = $\sqrt{14.3}$ = 3.8

The 75th percentile of the predicted demand counts is equal to E(1200) + 0.68 σ (1200) = 19.9 + 2.6 = 22.5. The 25th percentile is equal to E(1200) – 0.68 σ (1200) = 19.9 – 2.6 = 17.3. This means that with the probability 0.75 the demand counts will not exceed 22.5, and will not be smaller than 17.3. Moreover, with the 0.95 probability, the demand counts will be within the range of ± 2σ around expected demand, i.e., within the range from 12.4 to 27.4.

With this combination of demand counts at the three adjacent intervals, the expected value E(1200) = 19.9 appeared nearly equal to the deterministic prediction D(1200) = 20 for the 15-minute interval. Figure 5-4 illustrates this case with the range of uncertainty in demand prediction between 25th and 75th percentiles. Both expected demand and deterministically predicted demand are in the middle of the uncertainty range.

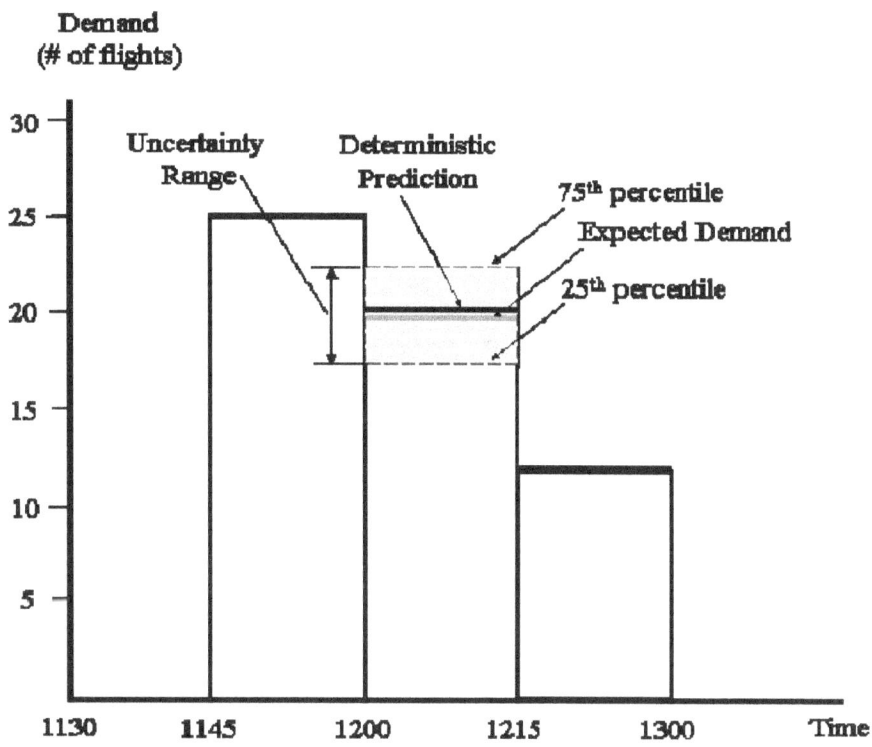

Figure 5-4 Case 1: Probabilistic vs. Deterministic Prediction

The calculations for Case 2 and Case 3 are similar to those for Case 1, and the results were presented in Table 5-3. They are shown graphically in Figures 5-5 and 5-6.

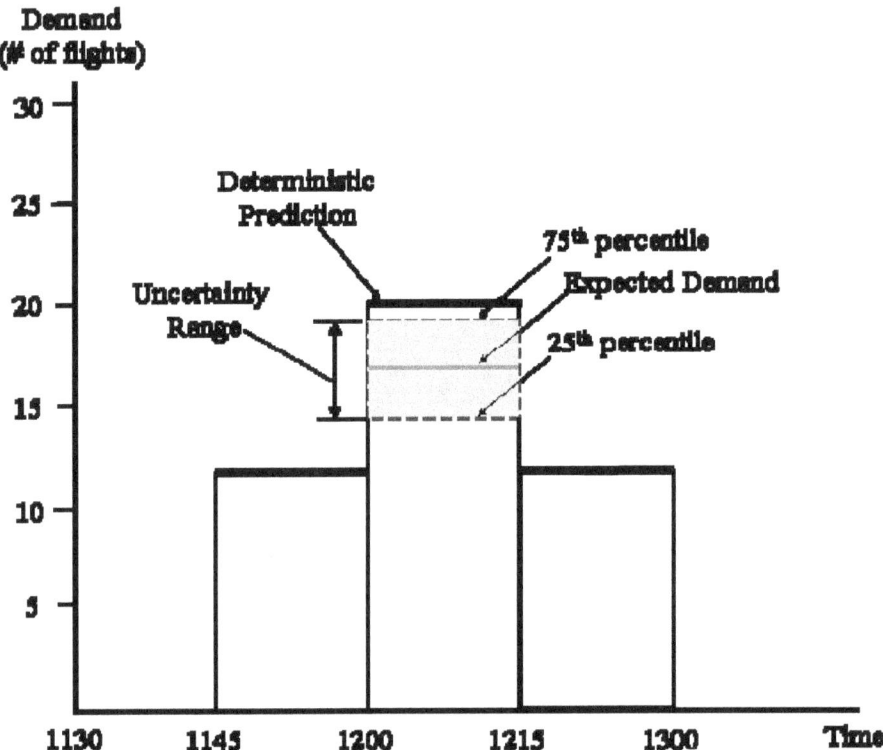

Figure 5-5 Case 2: Probabilistic vs. Deterministic Prediction

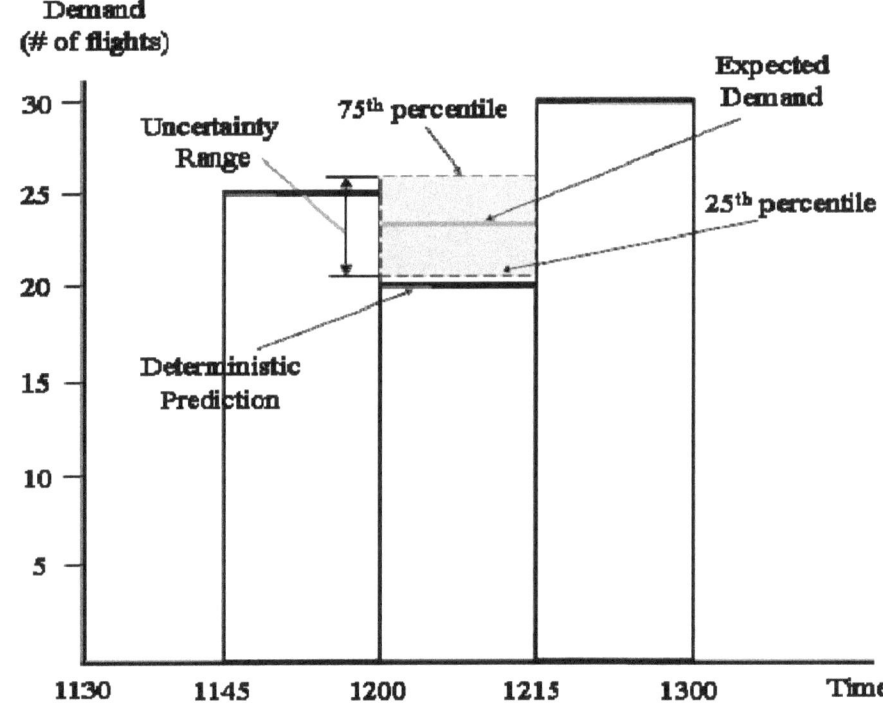

Figure 5-6 Case 3: Probabilistic vs. Deterministic Prediction

Note that, unlike Case 1, in both Case 2 and Case 3, the deterministic prediction of 20 flights is outside the uncertainty range between the 25th and 75th percentiles. In Case 2, the deterministic prediction is above the uncertainty range, but in Case 3 it is below (see Figures 5-5 and 5-6). It means that in both cases, the probability of traffic demand of 20 flights in the 1200 interval is smaller than 0.25.

These examples illustrate probabilistic demand prediction: it is measured by the expected value of demand counts and by the range of uncertainty around the expected value that covers the area within which predicted demand counts could have their values with a certain probability. The range is limited by the upper and lower bounds that correspond to probabilities for predicted demand to be higher and lower than a specified threshold. As for deterministic predictions, they make a significant contribution in calculating expected demand values and standard deviations that determine the level of predicted demand and parameters of prediction uncertainty (such as the range and probabilities), but the deterministic values themselves are not directly used as a probabilistic projection of demand. Instead, the expected demand counts, along with the uncertainty range covering the area around the expected demand with certain probability, determine the probabilistic demand predictions.

We can now give a general formulation of the algorithm for probabilistic demand predictions at 15-minute intervals based on characteristics of uncertainty in individual flights' ETA predictions and using deterministic demand count predictions at three consecutive 15-minute intervals. The algorithm processes deterministic aggregate 15-minute demand predictions for the interval of interest and for two closest adjacent intervals (the preceding and the following ones).

Consider a series of 15-minute intervals Δ_i

$$\Delta_i = [t_0 + i*15, t_0 + (i + 1)*15), \quad i = 1, 2, 3, \ldots$$

where t_0 is a starting time.

Let $D(\Delta_i)$ be a deterministic prediction of demand counts for interval Δ_i.

Then, according to equation (7), the following formula determines the expected value $E(\Delta_i)$ of predicted demand counts for interval Δ_i as a function of deterministic predictions for three consecutive intervals:

$$E(\Delta_i) = 0.26\, D(\Delta_{i-1}) + 0.57\, D(\Delta_i) + 0.17\, D(\Delta_{i+1}), \quad i = 2, 3, 4 \ldots \tag{13}$$

According to equation (9), the variance $\sigma^2(\Delta_i)$ of the predicted demand counts for interval Δ_i is equal to

$$\sigma^2(\Delta_i) = 0.22\, D(\Delta_{i-1}) + 0.35\, D(\Delta_i) + 0.15\, D(\Delta_{i+1}), \quad i = 2, 3, 4 \ldots \tag{14}$$

Equations (13) and (14) sequentially determine the basic values for probabilistic demand predictions at each 15-minute interval. In particular, knowledge of the standard deviations $\sigma(\Delta_i)$ for each 15-minute interval makes it possible to determine desired percentiles to quantify the range of uncertainty around expected values $E(\Delta_i)$. For example, the 75th percentile for interval Δ_i is equal to $E(\Delta_i) + 0.68\, \sigma(\Delta_i)$.

Consider a numerical example of deterministic 15-minute demand count predictions for 2.25 hours from 1200 to 1415. For this demand, expected values and variances were calculated by using equations (13) and (14).

Table 5-4 shows the calculation results along with deterministic predictions for each 15-minute interval.

Table 5-4 Example: Probabilistic vs. Deterministic Predictions

Time Interval	Deterministic Demand	Expected Demand	Expected Minus Deterministic Demand	Variance σ^2	Standard Deviation σ
1200 - 1215	20				
1215 - 1230	12	14.48	2.48	11.0	3.32
1230 - 1245	16	16.49	0.49	11.99	3.46
1245 - 1300	25	22.15	-2.85	15.57	3.95
1300 - 1315	22	21.25	-0.75	15.15	3.89
1315 - 1330	13	16.19	3.19	12.09	3.48
1330 - 1345	18	15.34	-2.66	10.66	3.26
1345 - 1400	10	14.63	4.63	11.21	3.35
1400 - 1415	25				

The table shows that the difference between deterministic and expected demand counts can be significant. For example, at 1345 – 1400 interval, the expected demand is more than 46% higher than the deterministic prediction. However, the standard deviations of probabilistic predictions do not vary much at different 15-minute intervals.

Figure 5-7 illustrates the probabilistic representation of traffic demand count predictions in the form of bar charts with the range of uncertainty at each interval restricted by 25th and 75th percentiles using numerical results from Table 5-4. At each bar, the red lines and the thick black horizontal lines show the predicted expected and deterministic demand values, respectively. At some 15-minute intervals, the expected values are close to deterministic ones (e.g., intervals starting at 1200, 1230 and 1300); at some intervals the deterministic predictions are much higher than expected demands and are outside the shown uncertainty ranges, which indicated a low probability for deterministic predictions (e.g., intervals starting at 1245, 1330 and 1400), and at some intervals the deterministic predictions are much lower than the expected values and also are outside the shown uncertainty range (see intervals starting at 1215, 1315 and 1345).

Figure 5-7 illustrates another important property of probabilistic demand prediction: the expected demands are smoothing the deterministic demand profile and are reducing fluctuations of predicted 15-minute demands along the time period that make it more stable.

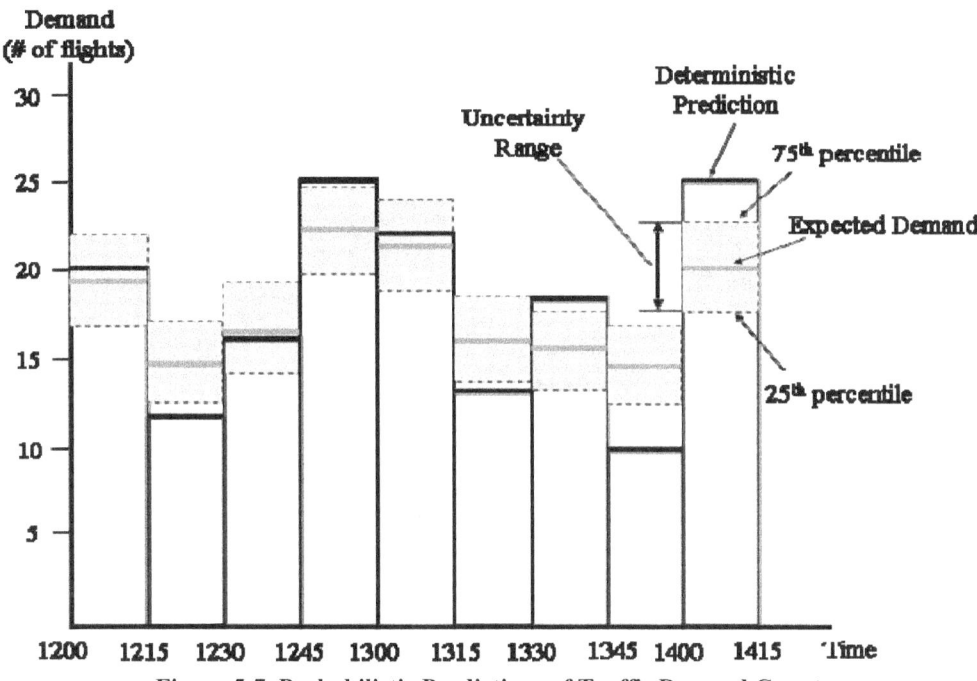

Figure 5-7 Probabilistic Predictions of Traffic Demand Counts

Another measure of interest is hourly demand, as TFM specialists might consider predictions for hourly demands during a several-hour period when making decisions on TMIs.

The example in Figure 5-7 shows that, although there might be significant differences between expected and deterministic predictions at 15-minute intervals, there is not much difference in the hourly counts. As Figure 5-7 covers only two and a quarter-hour period, we will illustrate it on the hourly counts within a sliding one-hour window that moves consecutively by 15 minutes. The results of the comparison are presented in Table 5-5.

Table 5-5 Hourly Demand Predictions: Expected vs. Deterministic Counts

One-hour Period	Deterministic Demand	Expected Demand	Expected Minus Deterministic Demand
1215 -1315	74	75	1
1230 – 1330	76	75	-1
1245 - 1345	78	74	-4
1300 - 1400	63	66	3

Table 5-5 illustrates that the hourly demand counts might not differ much. Further analysis will be needed to more thoroughly explore the differences between the hourly demand counts.

The above results on probabilistic prediction of traffic demand counts were obtained while using the following simplifying assumptions:

- Aircraft arrivals are statistically independent
- Binomial probability distribution for predicted demand counts is approximated by he normal distribution
- The probabilistic demand forecast provides an unbiased estimation of demand counts.

5.5 Comparison with Previous Work

In our 2007 report [2], we analyzed the accuracy of ETMS predictions of aggregate traffic demand counts and explored ways to improve prediction accuracy. The analysis was entirely based on 15-minue counts without usage of individual flight data.

The 15-minute arrival counts were examined for 9 airports for look-ahead times ranging from 0.5 to 6 hours. For those look-ahead times in the 1.25 – 2 hour range, the standard deviations ranged between 2.4 and 3.1 for those moderately busy airports with between 6 to 13 arrivals in 15 minutes [2].

For this report, a stochastic simulation was performed to roughly estimate how the errors in individual flight's predictions affected the accuracy of aggregate demand count predictions (see Appendix A of this report). The simulation model of Appendix A used an average demand of 7.5 flights in 15 minutes, a volume similar to MIA. In this simulation, the standard deviation of counts was found to range from 2.3 to 3.6, values consistent with what was observed in the ETMS data and reported in [2].

Along with the simulation, the analytical results were obtained in this study for estimating the accuracy of demand count predictions. In particular, equations (9) and (12) in Section 5.4 present a closed form approximation for the standard deviations of demand counts, based on the number of flights deterministically forecast in 15-minutes. If 7.5 flights are forecast in 15 minutes, the equations (9) and (12) yield standard deviations of 2.3 and 2.4, respectively, again, the values that are consistent with what was developed in the simulation model, and with what was observed in the ETMS data [2]

In 2007 [2], we also proposed, a linear regression model for 15-minute count predictions that made use of the count predictions in adjacent intervals [1, 2]. Linear regressions were performed on actual count data to determine coefficients for an adjacent-interval model of the form (applying to the example considered in previous section)

$$E(1200) = a\,D(1145) + b\,D(1200) + c\,D(1215) + k \tag{15}$$

In this earlier work, examination of the regression results revealed that reasonable values of the coefficients were a = 0.25, b = 0.55, c = 0.2 and k = 0, so that

$$E(1200) = 0.25\,D(1145) + 0.55\,D(1200) + 0.2\,D(1215) \tag{16}$$

Table 5-6 illustrates the relationship between these predictions and predictions just derived. They are practically similar.

Table 5-6 Coefficients in Airport Arrival Demand Models

	Equation (7)	Regression model (16) [1]
D(1145)	0.26	0.25
D(1200)	0.57	0.55
D(1215)	0.17	0.20

The similarity of these linear prediction models, obtained by two different methodologies, demonstrated the viability of proposed approach to probabilistic prediction of traffic demand.

Unlike the previous work that was limited by constructing the regression model for demand counts, this work developed a methodology that, in addition to regression model, provided the constructive way for quantifying the prediction uncertainty and its probabilistic properties.

6. Conclusion

ETMS currently makes its aggregate traffic demand predictions based on deterministic projections of traffic and neglects random errors in predictions. The purpose of this study was to

- Analyze the errors in time predictions for individual flights and characterize the accuracy of ETMS flight-by-flight prediction data
- Use these characteristics of uncertainty to determine the probability for an aircraft to be in a NAS element at any given time (e.g., to be in sector or to arrive at an airport)
- Relate the uncertainty in predictions for individual flights to uncertainty in predictions for 15-minute counts.

Analysis of accuracy of predictions for individual flights was separately conducted for flights' en route sector entry times and for flights' airport arrival times. In each case, comparative analyses of prediction errors were performed for different flights' status (active and proposed) as well as for different look-ahead times (from one to two hours and from two to three hours). As a result of this analysis of ETMS historical data, the characteristics of prediction errors were estimated.

For sector entry times, the results can be summarized as follows:

- Predictions for active flights are significantly more accurate than for proposed flights (that did not depart yet), with a much narrower range of uncertainty for the active flights: standard deviations of prediction errors are approximately 10 minutes for active flights, and 25 minutes for proposed flights.
- For active flights with 1 to 2 hour look-ahead time (LAT), the probability distributions of prediction errors appear symmetric with the average and median error close to zero, whereas for a longer LAT (from 2 to 3 hours) the distribution is asymmetric with a heavier right-hand tail and median error of 5 to 7 minutes, i.e., the distribution is biased with airborne flights tending to enter the sector earlier than predicted
- For proposed flights, the probability distributions of prediction errors are asymmetric with heavy left-hand tails that reflect a tendency for these flights to enter sectors later than predicted. There is no significant difference in prediction accuracy between a shorter and a longer LAT.

For flights' airport arrival times, the results were similar to those for sectors:

- Predictions for active flights are significantly more accurate than for proposed flights, with a much narrower range of uncertainty for the active flights. For active flights the prediction errors are within ± 15-minute range, whereas for the proposed flights the errors range from -45 to +25 minutes
- There is little difference in accuracy with shorter (1 to 2 hour) or longer (2 to 3 hour) LAT.
- For active flights, the probability distributions of prediction errors are nearly symmetric with zero average error.
- For proposed flights, the prediction errors are biased with median error of approximately -7 minutes that correspond to late arrivals. The probability distributions of prediction errors are asymmetric with heavier left-hand tails that reflect the tendency for proposed flights to arrive, on average, later than predicted.

A separate analysis was performed to estimate the distribution of the time-in-sector for flights. As expected, the flights' time-in-sector varies significantly by sector, with larger sectors requiring longer times. Typical time-in-sector ranged from 4 minutes (ZBW46) to 30 minutes (ZMP20). The analysis of accuracy of time-in-sector predictions showed that

- The prediction error for time-in-sector was noticeably smaller than for the sector entry time. The standard deviation of time-in-sector error is typically between 3 and 6 minutes (depending on the sector's size) whereas the standard deviation of the error of sector entry time is between 5 and 32 minutes (depending on flight status)
- There was not much difference in accuracy of time-in-sector prediction with respect to LAT or flight status (active or proposed),

The errors in predicting entry time and time-in-sector for individual flights were then used to determine probabilities that a flight would be in a sector at a particular moment in time given a deterministic prediction of flight's sector entry time. The analysis showed that the probability that a given flight would actually be in the sector during the time that it was predicted to be in the sector depended heavily on both the in-sector time and the flight status (active or proposed).

For example, when the predicted in-sector time was less than 20 minutes, 70 to 90 percent of the flights in the air were found to actually be in the sector at some point during the predicted interval, while less than 50% of flights on the ground were in the sector at some point during the predicted interval.

The results of analysis of uncertainty in individual flights predictions were used to develop probabilistic predictions of aggregate traffic demand counts. Analytical methods were used to obtain the probability distributions for traffic demand count predictions. The probabilistic characteristics of traffic demand counts for a specific time interval were determined from the set of flights that included not only the flights with ETAs within the time interval of interest but also the flights with ETAs in the immediate preceding and following adjacent intervals. The probability distributions characterize uncertainty in traffic demand predictions: they determine the most likely, expected values of traffic demand counts and the range of uncertainty around those expected values. The probabilistic demand predictions along with predicted capacities then can be used together to help determine the likelihood of congestion in NAS elements.

Appendix A Flight Event Predictions and 15-minute Counts: Simulation Experiments

In this section, we use simulation to explore how the uncertainty in predicting times of individual flight events transforms into uncertainty in predicting aggregate 15-minute traffic demand counts that are used as a basis for ETMS Monitor/Alert. It is intuitively clear that more accurate predictions for individual flights should lead to more accurate predictions of aggregate demand counts. For example, if the individual flight predictions were perfectly accurate, one would expect the 15-minute counts to be perfectly accurate.

We developed a spreadsheet-based stochastic simulation model to explore the relationship between flight-by-flight predictions and 15-minute counts. The simulation has two random elements. First, the initial schedule of the flights in the traffic flow was generated as a Poisson process with exponentially distributed time intervals between successive flights. Second, the random errors in predicted airport arrival time for individual flights were generated in accordance with the empirical distribution function obtained in Section 3 as a result of analysis of ETMS data (Figure 3-6).

This model serves three purposes:

- Relate flight-by-flight predictions to aggregate counts
- Provide a controlled environment for testing new prediction algorithms
- Assess the impacts of improvements in individual flight prediction accuracy on traffic demand count prediction accuracy.

The model was set up as follows (Figure A-1)

1. 250 flights were created, entering the system in a Poisson process with parameter $\lambda = 0.5$, corresponding to an average of 7.5 flights per 15-minute interval. Note: in the Poisson process, the average time interval between two consecutive flights entering the system is equal to $1/\lambda$. For $\lambda = 0.5$ min-1, the average inter-arrival time is equal to 2min, i.e., 7.5 flights per 15-minute on average.
2. Actual flight times between system entry and sector entry were assumed to be exactly 90 minutes.
3. The error in Predicted flight times was randomly drawn from the cumulative probability distribution in Figure 3-6. This error was combined with the actual flight times to produce simulated predicted flight times.
4. Predicted and actual 15-minute counts were calculated by applying the predicted and actual simulated sector entry times to the correct buckets.

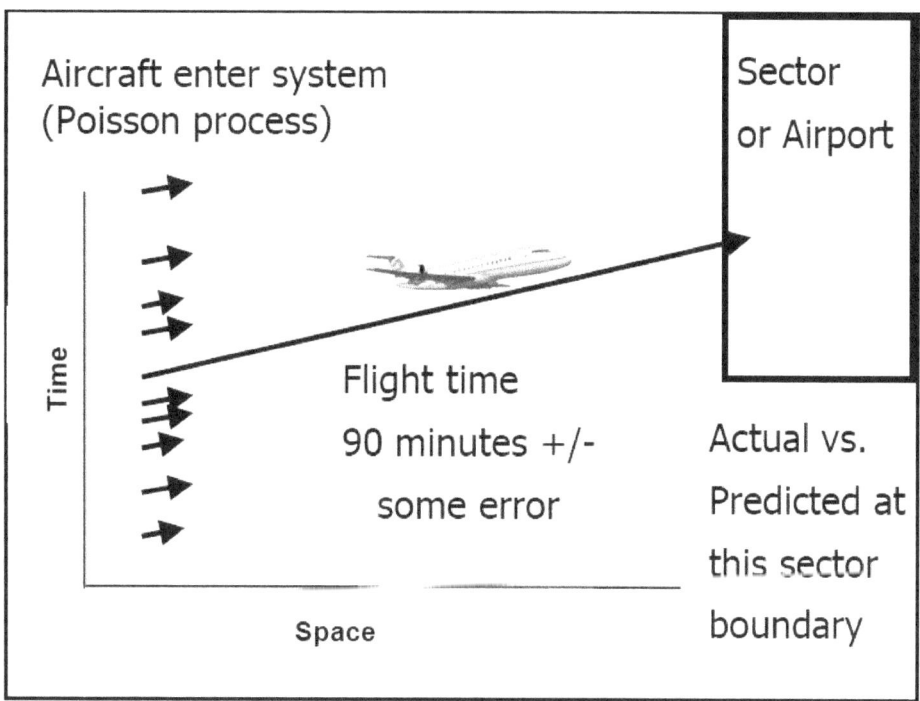

Figure A-1 Simulation Model

Results

First, we used the ETMS deterministic prediction algorithm to predict flights arriving in a 15-minute interval, and compared its accuracy with field data. The simulation was run 30 times, to produce a scatter plot comparing the standard deviation of predicted flight time error and the standard deviation of the error in the 15-minute counts (Figure A-2).

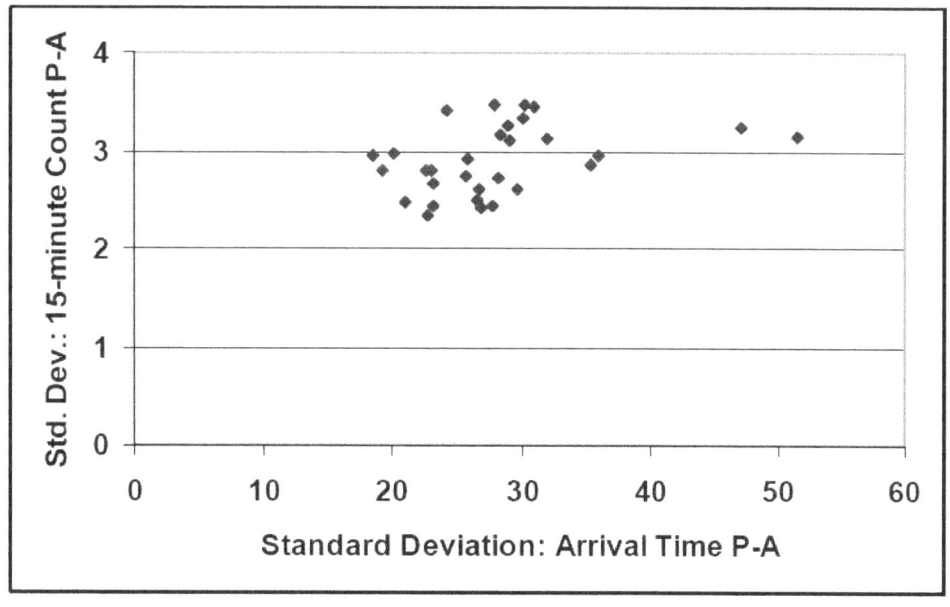

Figure A-2 Simulated Standard Deviations

A few observations from Figure A-2 are as follows:

- The arrival time standard deviation was typically between 20 and 30 minutes. This is consistent with the observed standard deviations of prediction errors for flights on the ground. This is not surprising, given that during the simulation the flight times were drawn from an empirical distribution of flight-time errors.
- The 15-minute counts were similar to what has been observed for airports with similar arrival volumes. For the airports of LAX, MIA, BOS and SFO, the standard deviation of count prediction error for LAT of 1.25 – 2 hours ranged from 2.4 to 3.1. (ref [1])
- The large standard deviations in flight time predictions (20 – 30 minutes) resulted in a relatively small (2.4 to 3.1) standard deviations in count predictions. This is presumably because the individual flight errors tended to cancel each other out.

The simulation was then used to assess the impact of improved arrival time predictions on count predictions. To do this, we "squeezed" the error distribution by a given factor, where 1 is the original distribution, and 0 would remove all of the error (Figure A-3).

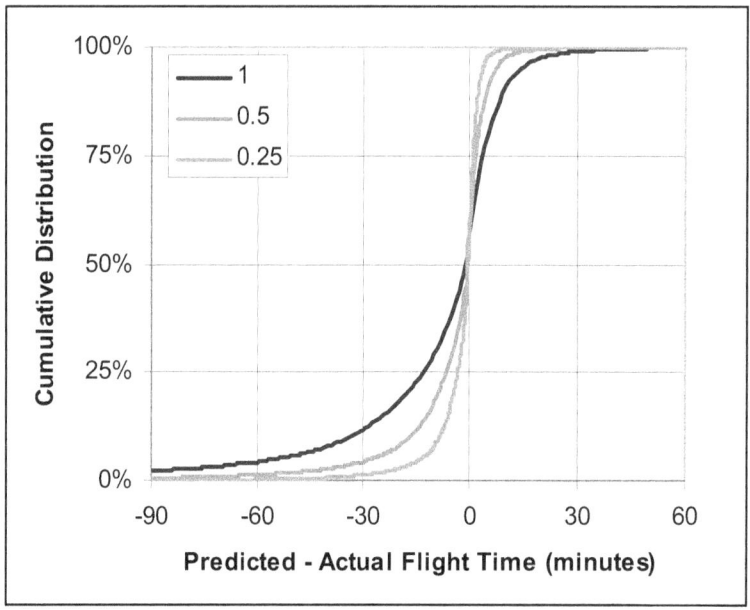

Figure A-3 Cumulative Probability Distribution of Errors, factors of 0.25, 0.5 and

The simulation was run 15 times, using factors of 0.1, 0.25, 0.5, and 0.75. The average of the standard deviation of the count error for the 15 runs was measured and plotted against the factors (Figure A-4). There is no significant reduction in the standard deviation of errors in counts (the Y axis) until the flight by flight error is reduced by at least 50% (the X axis).

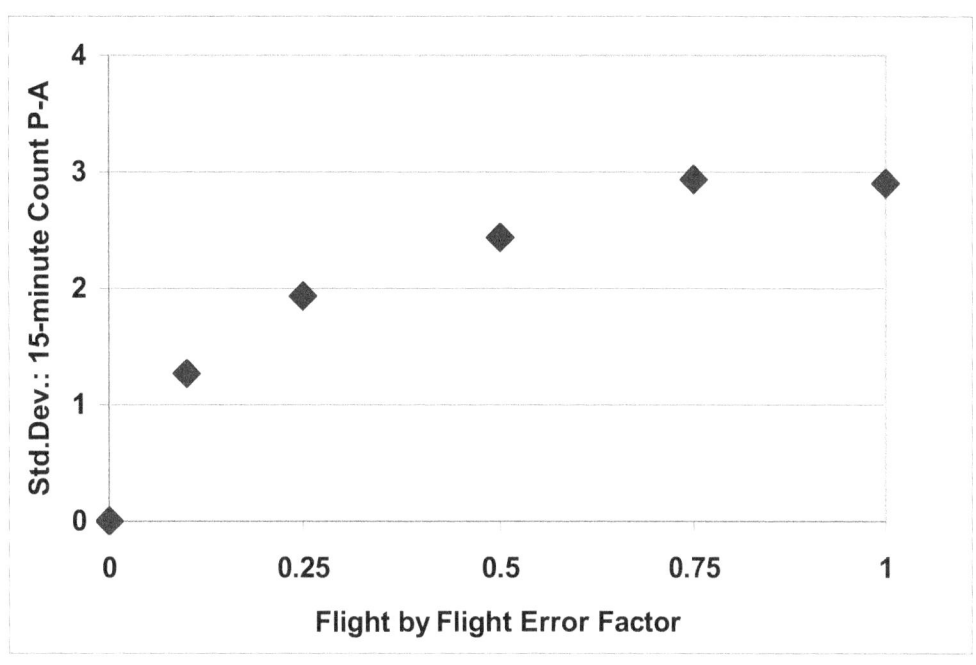

Figure A-4 Impact of Reducing Arrival Time Prediction Error

Appendix B References

[1] Gilbo, Eugene and Scott Smith "A New Model to Improve Aggregate Air Traffic Demand Predictions," Paper number 2007-6450, AIAA Guidance, Navigation, and Control Conference and Exhibit, Hilton Head, SC, 2007.

[2] Gilbo, Eugene and Scott Smith, "Reducing Uncertainty in ETMS Aggregate Traffic Demand Predictions," Volpe National Transportation Systems Center, Report no. VNTSC-CE-07-01, March 2007.

[3] Meyn, Larry A., "Probabilistic Methods for Air Traffic Demand Forecasting," Paper number 2002-4276, AIAA Guidance, Navigation, and Control Conference and Exhibit, Monterey, CA, August, 2002.

[4] Smith, Scott, and Eugene Gilbo, "Analysis of Uncertainty in ETMS Aggregate Demand Predictions", Volpe National Transportation Systems Center, Report no. VNTSC-ATMS-05-05, November 2005.

[5] Wanke, C., Song, L., Zobell, S., Greenbaum, D., and Mulgund, S., "Probabilistic Congestion Management," 6th USA/Europe Air Traffic Management R&D Seminar, Baltimore, MD, June 27-30, 2005.

[6] Wanke, C., Mulgund, S., Greenbaum, D., and Song, L., "Modeling Traffic Prediction Uncertainty for Traffic Flow Management Decision Support,", AIAA Guidance, Navigation, and Control Conference and Exhibit, Providence, RI, August, 2004.

www.ingramcontent.com/pod-product-compliance
Lightning Source LLC
Chambersburg PA
CBHW081901170526
45167CB00007B/3103